FLAVONOIDS, INFLAMMATION AND CANCER

FLAVONOIDS, INFLAMMATION AND CANCER

Hollie Swanson

University of Kentucky, USA

W🌀 World Scientific

NEW JERSEY · LONDON · SINGAPORE · BEIJING · SHANGHAI · HONG KONG · TAIPEI · CHENNAI · TOKYO

Published by

World Scientific Publishing Co. Pte. Ltd.

5 Toh Tuck Link, Singapore 596224

USA office: 27 Warren Street, Suite 401-402, Hackensack, NJ 07601

UK office: 57 Shelton Street, Covent Garden, London WC2H 9HE

Library of Congress Cataloging-in-Publication Data
Swanson, Hollie, author.
    Flavonoids, inflammation and cancer / Hollie Swanson.
        p. ; cm.
    Includes bibliographical references and index.
    ISBN 978-9814651936 (hardcover : alk. paper)
    I. Title.
    [DNLM: 1. Neoplasms--prevention & control. 2. Antineoplastic Agents, Phytogenic--
therapeutic use. 3. Flavonoids--pharmacology. 4. Inflammation--prevention & control. QZ 200]
    RC271.C5
    616.99'4061--dc23
                                        2015028010

British Library Cataloguing-in-Publication Data
A catalogue record for this book is available from the British Library.

Typeset by Stallion Press
Email: enquiries@stallionpress.com

Contents

Introduction to Flavonoids and Chemoprevention*

1

The use of plant materials to treat human diseases can be found through-out history and is documented in the ancient texts of China, Egypt, and India. In ancient China, the best recognized, earliest work was authored in approximately 2700 BC by Emperor Shen Nung, who is often referred to as the father of Chinese medicine.[1] He is credited with tasting 365 herbs and classifying them in part, on their relative toxicity. Many Chinese herbal medicines that are currently used can be traced to the work of Emperor Nung. In ancient India, the system of medicine known as Ayurveda, is thought to have originated between 1500 and 2000 BC.[2] Ayurveda texts cite 6,000 medicines which are primarily plant-based to be used for the treatment of a variety of ailments and diseases. Traditional Ayurvedic medicines are polyherbal and contain combined mixtures of plant extracts that are thought to interact synergistically. Approximately 15,000 medicinal plants in India have been documented, but the full identification of medicinal plants and their potential therapeutic value is still thought to be incomplete. Ongoing research efforts focused on iden-tifying active pharmacological components of these herbal mixtures have discovered the integral role that flavonoids play in these ancient recipes.

Another important source of traditional medicine can be found by exploring the history of Egypt. The earliest text from ancient Egypt that describes 829 prescriptions for disease treatment is the Ebers papyrus and is thought to have been written in 1500 BC.[3] These primeval treatments that were used by the Egyptians of that time were obtained primarily from plants and trees, but also utilized other substances that were obtained from

---

*Abbreviations: 8-OHdG, 8 hydroxy-2′deoxyguanosinde; EGCG, (−)-epigallocatechin-3-gallate; NSAID, non-steroidal anti-inflammatory drug; UV, ultraviolet light.

animal sources including pig eyes. The prepared remedies were taken orally or were applied externally as poultices or unguents. The Egyptians also developed crude methods for delivering vaporized medicine. One method required the use of a double pot. The liquid remedy was poured into a hot pot. A second pot that contained a hole was then placed on top. A straw was inserted into the hole and the patient breathed the vaporized substance through the straw. A second approach involved laying the herbs onto hot bricks. This would release the alkaloid-containing active ingredients of the plant as vapor. Our current labeling of prescription vials is thought to originate from the practice of Egyptian physicians who would often write their prescriptions on small labeled containers which resembled small cylindrical pottery vases. Like Ayurvedic medicines, those of ancient Egypt often contained complex mixtures of herbs. The transfer of knowledge of these ancient medicines was aided by both extensive trade routes and religious practices including pilgrimages and the Crusades. Many of the European monasteries served as repositories of the medical knowledge obtained from these ancient sources where their therapeutic approaches were implemented and refined.

Our current interest in these plant-based medicines that have rich origins in China, India, and Egypt stems from our desire to procure their health benefits in a manner that is both safe and effective. Consumption of herbal medicines is growing worldwide as they are used not only as therapies to cure specific disease states but also as supplements taken with their prescribed "western" medicines. With the advent of modern pharmacological practice, active research has focused on identifying the bioactive ingredients of these traditional herbal preparations, their pharmacological mechanisms of action, their efficacy, and their potential adverse effects. The primary bioactive chemical constituents of plants used for medicinal properties are flavonoids and proanthocyanidines (oligomers of flavonoids), tannins, terpenoids, resins, lignans, alkaloids, forocoumarines, and naphthodianthrones.[4] In many cases, the pharmacological activity of the traditional medicines varies considerably and thus these medicines are used to treat a wide range of disease states. For example, flavonoids, terpenoids, and resins are highly valued for their anti-diarrheal properties, whereas alkaloids and terpenoids hold promise for their potential for combating malaria.[5,6] Our scientific investigations probing the therapeutic benefits of flavonoids can be traced to experiments on vitamin "P," a flavonoid preparation isolated from citrus fruit which was found to protect against radiation-induced injury and lethality.[7] This early work provided the foundation for the current investigations into the role of flavonoids in providing protection against chronic disease states including diabetes, cardiovascular diseases, and cancer, protection which is often attributed to their anti-oxidant and/or anti-inflammatory activities.[8,9] Over the past 20 years, typical experimental paradigms have involved administering varying concentrations of isolated flavonoids to either cultured cancer cells or to laboratory animals, followed by the evaluation of cancer-relevant endpoints. Upon accumulation of sufficient evidence supporting their ability to inhibit key biological markers in the absence of adverse effects, clinical phase I trials in human subjects are then initiated. Given that flavonoids are present in relatively high concentrations in many foods and beverages, additional evidence pertaining to their potential health benefits can also be accrued using epidemiological approaches. However, as we attempt to extrapolate findings obtained using isolated, individual flavonoids to develop effective therapeutic and/or chemopreventive agents, we should be mindful of a major tenet of the ancient healers, that individual components may interact synergistically and thus as stated by Aristotle, it may be that "the whole is greater than the sum."

As a first step toward understanding how flavonoids may exert their chemopreventive properties, we turn towards examining their chemical structures (Fig. 1.1). The basic chemical structure of flavonoids has been described as C6-C3-C6; that is, two aromatic rings that are connected by a three carbon bridge.[8] The classification of flavonoids is based on their chemical structures which vary in the number of double bonds, the number of oxygen-containing substituents and their relative positions. There are six classes of flavonoids: (1) flavanones, (2) flavonols, (3) flavones, (4) isoflavones, (5) flavan-3-ols, and (6) anthocyanidins. The basic structure of these flavonoid classes is shown in Fig. 1.1. In plants and upon consumption of plant products, these basic structures often undergo further modification which involves their hydroxylation, methoxylation, or O-glycosylation of the hydroxyl groups.[10] In addition, the carbon atom may be glycosylated and additional alkyl groups may be covalently attached. The presence of isomeric forms of flavonoid aglycones and the

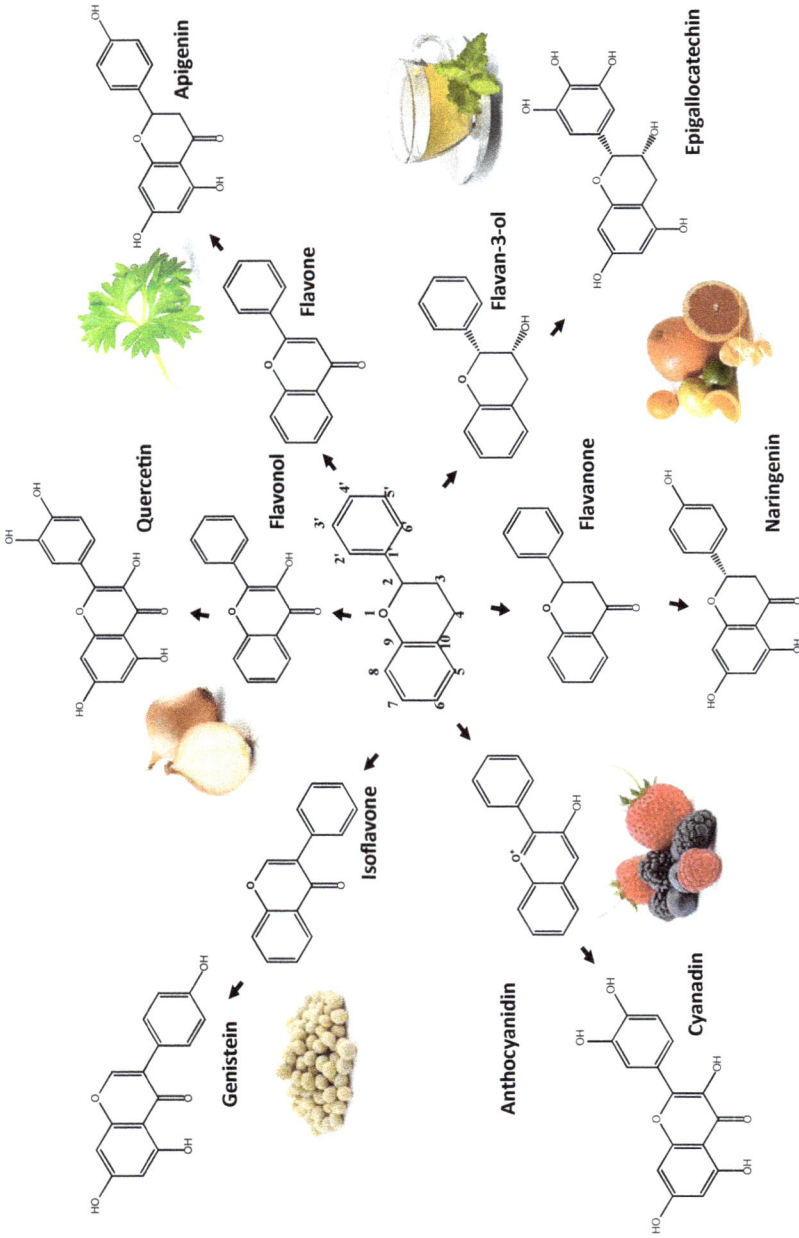

Fig. 1.1. A scheme of the six classes of flavonoids and representative flavonoids commonly found in foods and beverages.

varying patterns of glycosylation often present challenges in the ability to accurately identify and quantify flavonoids in plant-derived food. Further, metabolism by the host organism and its microbiome results in additional modifications that typically include formation of conjugates (i.e., methyl, sulfates, and glucuronides) that often alter the biological property of the flavonoid. Many of these flavonoids harbor intensely studied anti-oxidant, anti-inflammatory, and anti-tumor activities as will be discussed in more detail in subsequent chapters.

Human consumption of flavonoids varies depending on global locale, age, gender, socio-economics and dietary preferences. Estimates obtained from a number of studies indicate that the average daily intake in the majority of the global population ranges from approximately 180 to 200 mg.[11–14] Tea, fruit, and wine are the major food sources of flavonoids of most population groups. Amongst the most highly consumed flavonoid subclasses are the flavanones and flavano-3-ols.[13,15] However, the flavonoid levels as well as the relative amounts of the individual flavonoid compounds within dietary sources can vary significantly depending on the cultivar of plant species, their growth conditions, and the preparation and storage of the food product. Flavonoids representing each subclass and their common food sources are shown in Fig. 1.1.

Flavanones. Some of the best characterized flavonones are hesperidin and naringenin which are highly abundant in citrus fruits.[16] Naringenin is well known for its bitter taste and its ability to impair drug metabolism. Thus, consumption of foods high in naringenin, such as grapefruit juice, is often involved in drug-supplement interactions that arise from the inhibition of drug clearance and inappropriately high drug plasma levels. Other flavanones of interest include pinocembrin (5,7-dihydroxyflavanone) found in plants such as Eucalyptus[17] and aglyca which is thought to contribute to the anti-inflammatory properties of willow bark.[18]

Flavonols. The most important contributors of flavonols are thought to be quercetin, kaempferol, myricetin and isorhamnetin which are found in high concentrations in berries.[8,19] Quercetin is also abundant in the skins of apples and onions.[20] Rich sources of kaempferol include tomatoes,[21,22] carrots, tea,[23] currants,[19] onions,[24] cucumber, mustard, coriander, watercress,[25] and broccoli.[16] Other flavonols of interest include wogonin (5,7-dihydroxy-8-methoxyflavone), a constituent of *S. baicalensis*, a perennial herb that is used in traditional Chinese medicine,[26] as well as rutin and nobiletin found in citrus fruits.[27]

Flavones. Well-characterized flavones include apigenin and luteolin which are found in relatively high concentrations in food sources such as onions, celery, and pistachio,[28,29] aromatics such as lavender,[30] and folk medicine such as burr parsley.[31] A derivative of apigenin, 5-O-caffeoylquinic acid, has recently been identified as an abundant compound with high antioxidant properties in Roman chamomile which is also used in traditional medicines.[32]

Isoflavones. The diphenolic isoflavones are found in the highest concentrations in legumes, in particular, soybeans, red clover, and kudzu root.[33] Isoflavones contribute approximately 0.6% of the daily flavonoid intake in the United States. Of the isoflavones, the most highly consumed include daidzein and genistein which are present in soybeans and are consumed in products such as tofu, soy milk, and numerous meatless foodstuffs.

Flavan-3-ols. The average intake of flavan-3-ol intake ranges from approximately 150 to 450 mg/day.[13,34] Major food sources of flavan-3-ols are apples and other pome fruits, stone fruits, berries, wine, and tea. Nuts, in particular pecans and pistachios, are also good sources of flavan-3-ols. Amongst the most highly consumed flavan-3-ols are catechin gallate, epicatechin gallate, (–) 3 epigallocatechin-3-gallate and gallocatechin.[16]

Anthocyanidins. The brightly colored hues of red, blue, and purple fruits and vegetables arise from the presence of anthocyanidins. The most commonly occurring anthocyanidins are cyanidin, delphinidin, malvidin, pelargonidin, peonidin and petunidin. Human consumption of anthocyanidins typically ranges from 12 to 65 mg/day.[35] Important food sources of anthocyanidins include berries, stone fruits, eggplants, black beans, red cabbage, red onions, and red radishes. While anthocyanidins are also present in wine and grape juice, their levels are considerably lower as compared to that found in fresh blackberries, blueberries, and raspberries (i.e., 10–14 mg/100 g versus 245–390 mg/100g).

Overview of chemoprevention. Cancer chemoprevention has been defined as "pharmacological intervention with synthetic or naturally occurring compounds that may prevent, inhibit, or reverse carcinogenesis or prevent the development of invasive cancer."[36] The early experimental work on chemoprevention focused largely on the ability of an administered dietary or chemical agent to inhibit chemically induced carcinogenesis in either a laboratory rat or mouse. Mechanistically, these agents were often found to either inhibit the enzymes involved in the metabolic activation of administered carcinogens or induce the expression of enzymes responsible for "detoxifying" the carcinogen.[37] Since those early years of discovery, our understanding of the carcinogenic process has evolved and the experimental animal models that we use have become more refined to more appropriately mirror events that occur during the development and progression of human cancers. Accordingly, our understanding of the chemopreventive effects of naturally occurring compounds has deepened and revealed the complexity of their potential actions. Nonetheless, the overarching goal remains constant: to identify the ideal chemopreventive agent which is inexpensive, easily administered, effective, and poses limited side effects. Individuals who are determined to be at high risk of developing neoplasia, either using genetic profiling or by detecting the presence of dysplasia, would then be administered this ideal chemopreventive agent. Given these stringent requirements and the fact that chemopreventive regimens are typically administered for extensive time periods, food and natural products represent excellent sources for identifying ideal chemopreventive agents.[36]

Chemoprevention using pharmacological agents. Extensive study of molecular targeted approaches have culminated in the U.S. Food and Drug Administration (FDA) approval of anti-inflammatory agents and anti-estrogens to inhibit the risk of developing colorectal and breast cancers, respectively.[38] The initial study supporting the idea that inhibiting inflammation would protect against the development of colorectal adenomas involved only 22 patients treated with the anti-inflammatory drug, sulindac.[39] This work not only paved the way for the use of these non-steroidal anti-inflammatory drugs (NSAIDs) for chemopreventive purposes, but also led to our understanding of the extent to which the inflammatory response contributes to the carcinogenic process. However, it is not yet clear whether NSAIDS would be similarly effective for chemopreventive strategies for other cancers as results obtained thus far have been mixed.[40] The chemopreventive use of anti-inflammatory agents to prevent gastrointestinal and steroid-responsive cancers will be discussed further in Chapter 4.

The use of anti-estrogens as chemopreventive agents capable of inhibiting breast cancer can be traced to the proposed use of tamoxifen as an oral contraceptive in the early 1960s.[41] At around the same time, it was realized that removing the ovaries of mice susceptible to developing mammary tumors, reduced tumor incidence. Due in part to the lack of interest in breast cancer therapies, ICI46, 474, now known as tamoxifen, was developed and marketed as a drug with dual use, for inducing ovulation and for treating breast cancer. As the need for breast cancer treatments became more apparent and with the realization that tamoxifen had a relatively high safety profile, its potential use as a chemopreventive agent to inhibit breast cancer became more popular. The use of anti-estrogens as chemopreventive agents for breast cancer will be discussed in more detail in Chapter 5. Given the emerging status of the estrogen receptors in tumorigenesis affecting tissues such as the lung[42] and colon,[43] it is highly likely that the use of estrogen receptor-targeting chemopreventive agents will rapidly expand to include treatment of tumors of non-reproductive tissues.

Chemoprevention using fruits, vegetables, and tea. Edible plants that have been ascribed chemopreventive properties include the cruciferous vegetables, such as broccoli, cabbage, and cauliflower. Epidemiological evidence as well as that generated using animal models of carcinogenesis indicates that high consumption of cruciferous vegetables plays a protective role in decreasing the incidences of colorectal, renal, bladder, lung, and prostate cancers.[36,44] Cruciferous vegetables are consumed primarily in Asia, Europe, and North America. The estimated consumption of broccoli in these countries ranges from 3 to 6.8 g/capita/day. Key chemical constituents that are thought to mediate the chemopreventive activities of cruciferous vegetables are glucosinolates, in particular, isothiocyanate, and their hydrolysis products. In experimental models of carcinogenesis, these agents have been shown to hinder the metabolic activation of chemical carcinogens such as nitrosamine 4-(methylnitrosamino)-1-(3-pyridyl)-1 butanone, a constituent of tobacco smoke that can cause lung cancer.[45] Depending on the food source, dietary intake of up to 3 mg isothiocyanate per gram of vegetable can be readily consumed. The underlying chemopreventive effects of isothiocyanates involve inhibition of cytochrome P450 enzymes, activation of the phase II conjugating enzymes, and induction of tumor cell apoptosis. In addition, isothiocyanates inhibit cellular transporters involved in mediating "phase III" metabolism. An isothiocyanate with particularly potent chemopreventive properties, sulforaphane is found in relatively high levels in broccoli.[46] However, sulforaphane, is chemically unstable and has been shown to provoke gastric irritation when administered to rats in its pure form.[44] These adverse properties have limited further investigations into the development of sulforaphane as a chemopreventive agent.

Additional studies focused on broccoli have revealed that several broccoli constituents, in addition to isothiocyanate, are capable of exerting chemopreventive effects.[44,47,48] The primary components of broccoli are glucosinolates that exist in approximately 120 different aglycone forms. The individual components of glucosinolates, as well as of other constituents of broccoli, vary considerably amongst different cultivars, the plant part, and the environmental growth conditions. Glucosinolates undergo extensive metabolic breakdown. At low pH, like that encountered in the human gut (pH 6–7), glucosinolates are hydrolyzed into stable isothiocyanates which undergo further lysis to form indole-3-derivatives. These derivatives, including 5,11-dihydroindolo[3,2-b]-carbazole, can activate the aryl hydrocarbon receptor which is an important regulator of drug metabolizing enzymes. As a consequence, the indole-3-derivatives, like indole-3-carbinol and its condensation products (i.e., 3,3′ diindolylmethane), can exert chemopreventive effects, in part, due to their ability (via activation of the aryl hydrocarbon receptor) to accelerate the degradation of endogenous estrogen and inhibit its tumor-promoting effects.[49] In addition, indole-3 carbinol, as well as several isothiocyanates, has been shown to induce cell cycle arrest and apoptosis, inhibit angiogenesis, modify DNA methylation, and inhibit the actions of certain viruses and bacteria associated with human cancers (i.e., human papilloma virus and *Helicobacter pylori*). These activities, in their composite, are considered to be anti-tumorigenic. Despite these promising results, the use of indole-3-carbinol as a chemo-preventive agent is accompanied with a cautionary warning as some studies have reported that its administration can result in pro-tumorigenic actions. For example, indole-3 carbinol treatment has been found to enhance the formation of chemically induced liver foci[50] and its condensation product, 3,3′ diindolylmethane, can enhance the expression of estrogen-responsive genes *in vivo*.[51] Resolution of the "pro" versus "anti" activities of isothio-cyantes and indole-3-carbinol in the human population requires accurate assessment of the appropriate dose that an individual should receive. The possibility exists that these opposing effects of indole-3-carbinol and other putative chemopreventive agents are dependent on dose and are a function of a U-shaped dose response curve where at lower doses are "anti-tumorigenic", but at higher doses are "pro-tumorigenic." To resolve this issue, it is essential to determine the tissue levels that are present in an individual following their consumption of a given dietary substance such as cruciferous vegetables. Recent advances in detection methodologies have addressed this uncertainty and report that urinary levels of 3,3′ diindolyl-methane can be detected in human subjects following their ingestion of Brussels sprouts at levels that are higher than those detected in individuals who had ingested cabbage.[52]

Other broccoli constituents with prescribed chemopreventive proper-ties are ascorbic acid (vitamin C), phenols, flavonoids, xanthophylls, carotenoids, tocopherols, polyunsaturated fats, and chlorophyll. The major flavonoids found in broccoli are quercetin, kaempferol, and isor-hamnetin, all of which exert anti-oxidant activities. Total flavonoid con-tent of broccoli has been found to be higher in the leaf tissue as compared to the florets. Flavonoid concentrations in broccoli, as expressed as mil-ligrams of catechin hydrate equivalent per 100 grams of dry weight (mg CE/100 g), has been reported to range from 200 to 800 mg CE/100 g. The bioavailability of these flavonoids in broccoli is thought to be poor, but more research is needed to understand how these flavonoids, as compo-nents of our daily food, when ingested as supplements or when taken in their individual forms, are metabolized and absorbed. The potential adverse effects of excessive consumption of broccoli or broccoli-based supplements are indicated by evidence of genotoxicity observed using both the Ames assay and laboratory animals.

The culmination of the pre-clinical studies investigating the chemo-preventive effects of cruciferase vegetables is represented by two studies performed in human subjects that were designed to determine whether the active components of cruciferous vegetables could minimize human exposure to chemical carcinogens present in either the diet or the environment.[53,54] First, a randomized, placebo-controlled chemopreventive trial performed in China used a broccoli sprout-infused beverage to deliver high levels (400 μM) of glucoraphanin, a metabolic precursor of sulforaphane.[53] This population has been found to be exposed to relatively high levels of aflatoxin, a genotoxic agent that is formed during improper storage of wheat, peanuts, and other crops. High exposure to aflatoxin can contribute to the development of hepatic tumors. In this study, exposure to aflatoxin was determined by measuring the urinary levels of aflatoxin-DNA adducts (aflatoxin-N7-guanine). An inverse relationship between urinary levels of sulforaphane metabolites and aflatoxin-DNA adducts was observed in the subjects who had ingested the broccoli sprout beverage. A similar study, again using a broccoli sprout beverage was performed to examine its effects on ameliorating exposures to airborne pollutants.[54] The human subjects ingested broccoli sprout beverages containing either 800 μM glucoraphanin, 150 μM sulforaphane, or both. The results from this short-term study indicated that ingestion of either glucoraphanin or sulforaphane can be effective at enhancing the elimination of acrolein, crotonaldehyde, or benzene. Adverse effects associated with broccoli sprout beverages appear to be limited to indigestion and light vomiting experienced by a few individuals. Clinical studies focused on determining the impact of broccoli sprout extracts on breast cancer proliferation, advanced pancreatic cancer, and lung cancer are currently ongoing (Clinicaltrials.gov, accessed March 12, 2015).

Fruits, in particular berries and grapes, have also been shown to have chemopreventive properties.[55] Seminal work investigating the chemopreventive properties associated with berries initiated with the study of individual components, ellagic acid and phenylethyl isothiocyanate. The primary laboratory model used to study the chemopreventive effects of these agents was a model of esophageal cancer in rats initiated by

N-nitrosomethylbenzylamine, a potent carcinogen present in tobacco smoke. While ellagic acid was found to be an effective chemopreventive agent in the rat model, its poor bioavailability limited further study in human subjects. In addition, using sophisticated chemical analyses, it became evident that berries contained multiple (putative) chemopreventive agents including not only ellagic acid, but also calcium, β-sitosterol, quercetin, and several anthocyanins, such as cyanidin-3-O-rutinoside. With this in mind, efforts then turned to the use of food sources that contained high levels of ellagic acid: blackberries, red raspberries, and strawberries. The highest concentrations of ellagic acid was found in the pulp and seeds of these fruits with considerably lesser amounts in the juice. By lyophilizing the berries, the concentrations of ellagic acid and other substances could be concentrated. The initial studies again used the rat model of chemically induced esophageal cancer and dietary levels of berry powder that were comparable to 2–4 cups of fresh berries per day. The cumulative results of these studies indicated that this berry preparation inhibited the formation of tumors, but not in a dose-related manner. In some studies, the lower (5%), but not higher (10%) doses of berry powder were found to be inhibitory. Anthocyanin-rich fractions of the lyophilized berry extracts also proved to be effective at inhibiting tumor formation. In addition to ellagic acid and anthocyanins, ellagitannins were also found to be important for eliciting chemopreventive effects. With respect to chemopreventive mechanisms, inhibition of proliferation, apoptosis, and oxidative stress are thought to be involved.

Studies testing the effects of lyophilized berry extracts in human subjects include a phase I clinical trial to examine the pharmacokinetics of anthocyanins and ellagic acid. Uptake of ellagic acid and anthocyanins from the administered berry extracts was found to be less than 1% of the administered dose.[56] Clinical trials that have been performed to determine whether components of berries may slow the progression of human cancers include a pilot study performed in patients determined to be at high risk for developing oral cancer.[57] Here, a bioadhesive that contained 10% freeze dried blackberries was applied to premalignant lesions found within their oral cavities. The bio-adhesive (0.5 grams) was administered four times per day for six weeks and biopsies were obtained prior to and after the six-week treatment period. No adverse effects were reported.

In biopsies obtained from some of the patients, changes in the expression levels of genes involved in apoptosis and terminal differentiation were increased. Additional clinical trials have been performed to determine the safety and effectiveness of lyophilized berry extracts on the progression of GI cancers (i.e., esophageal adenocarcinoma and colorectal adenomas). These will be discussed in more detail in Chapter 4.

Tea is one of the most popular chemopreventive agents.[58,59] Tea, the water soluble extract of the dry leaves of *Camellia sinesis*, is consumed as either green, oolong, or black tea, which varies with respect to the extent of their fermentation. Green tea is consumed as the non-fermented form, oolong tea is semi-fermented, and black tea is fully fermented. Worldwide, black tea represents the majority of tea produced. The major polyphenols present in tea are catechins in green tea and the aflavins and the arubigins in black tea. In green tea, 50 to 80% of the total catechins present is in the form of (−)-epigallocatechin-3-gallate (EGCG). Other flavonoids found in tea in lesser amounts are the flavonols, quercetin, kaempferol, and myricitin and their glycosides. Flavonoids represent approximately 0.5–2.5% of infused tea extracts. The water-extracted solids in tea contain approximately 2.5% caffeine which is present in similar concentrations in both green and black tea. Interestingly, the tea catechins (i.e., EGCG) have been found to be relatively unstable during cell culture conditions typically used in the laboratory. They typically undergo reactions such as oxidation and multimerization. These reactions do not appear to occur *in vivo* and thus may account for differences obtained when comparing *in vitro* to *in vivo* results.

Tea, tea extracts, and pure EGCG have been shown to inhibit chemically induced cancers of the lung, skin, and colon in laboratory animals.[58] Tea has also been shown to exert protection against tumors formed in genetically engineered mice and the growth of explanted human tumors. The underlying chemopreventive mechanisms appear to involve upregulation of the cellular anti-oxidant defenses pathways, inhibition of inflammation, inhibition of angiogenesis, and inhibition of tumor cell survival. In addition, EGCG has been shown to directly interact with proteins involved in tumor progression such as the heat shock protein of 90 kDa, HSP90.[60]

Human intervention studies performed thus far indicate that green tea and/or EGCG may be effective at inhibiting the development of tumors at the early, but not later, stages of cancer progression.[58] The studies that have investigated the effects of catechins have used either tea (typically green tea) or extracts prepared from tea. A preparation of extracts commonly used is Polyphenon E®. Polyphenon E® is a standardized extraction of green tea leaves, *Camellia sinensis*.[61] It is formulated as a capsule that contains approximately 70% EGCG and at least ten different catechins, and was developed by a partnership between the U.S. National Cancer Institute and Mitsui Norin Co, Ltd. It is approved by the FDA as a botanical drug and has been tested for its chemopreventive effects in more than 20 phase I and phase II clinical trials. Clinical intervention trials that have examined the effects of Polyphenon E® include a study involving women with cervical cancer.[62] In this phase II clinical trial, 98 women diagnosed with human papillomavirus (HPV) infection and low-grade cervical intraepithelial neoplasia were administered either Polyphenon E® capsules (equivalent to 800 mg EGCG) or a placebo control for four months. The average urinary (−)-epigallocatechin levels were 652 ng/ml in the treated versus 59 ng/ml in the controls. The reported adverse effects were minimal. There was no significant impact of Polyphenon E® treatment on either HPV infection status or cervical cancer progression. The impact of green tea and green tea extracts on colon, breast, and prostate cancer will be discussed in Chapters 4 and 5.

A similar preparation called Polyphenone 70A, a green tea powder containing 55.9% EGCG, has been used to test the impact of tea catechins on the progression of oral cancer.[63] The advantage of this approach is that the catechins can be directly administered to the tissue of interest and in this manner may circumvent the problems associated with their poor bioavailability. In this study, seven patients who were diagnosed with oral field cancerization but with no evidence of oral squamous cell carcinomas participated. The Polyphenone 70A powder was incorporated into a mouthwash that contained 800 mg EGCG. The participants rinsed with the mouthwash for two minutes, once per day for seven days. While no measureable levels of EGCG in the blood and saliva could be detected prior to the onset of the study, after seven days, EGCG was detectable in the saliva, but not the blood, of all participants. The exfoliated oral epithelial cells were examined for markers of proliferation and inflammation.

As compared to their baseline values, decreases in both markers were observed in some of the patients following their exposure to EGCG.

The impact of green tea catchins on inflammation has also been examined using a model of UV-light induced skin inflammation.[64] Here, oral supplements (taken as three capsules) consisting of a total of 1,350 mg tea, 540 mg green tea catechins, and 217.8 mg EGCG combined with vitamin C (50 mg) was administered orally on a daily basis for 12 weeks. Vitamin C was added in an attempt to protect the green tea extract from degrading in the lumen of the gastrointestinal tract. The average urinary excretion of (−)-epicatechin and (−)-epicatechin-3-O-gallate was 6.1% and 7.1%, respectively, of the administered dose in the 16 patients. The subjects were exposed to UV light prior to and following the administration of the green tea catechins to produce sunburn-associated inflammation (i.e., erythema). The results from this trial indicated that oral administration of green tea extracts can significantly decrease the development of UV-induced erythema as well as reduce the amount of arachidonic acid metabolites (associated with the inflammatory response) formed.

Others have questioned whether the impact of flavonoids, such as those present in green tea, would be more effective in individuals who smoke tobacco products and are at high risk of developing cancer. Using a phase II randomized controlled study design, Hakim *et al.* questioned whether consumption of tea would impact oxidative damage in smokers.[65] 143 heavy smokers were randomly assigned to drink four cups of either water, decaffeinated green tea, or decaffeinated black tea per day for four months. The amount of EGCG consumed was estimated to be 14.56 mg/day for the black tea drinkers and 143.84 mg/day for the green tea drinkers. Total flavonol glycosides consumed per day was 28.6 mg (black tea) and 30.08 mg (green tea). Green tea consumption was found to significantly reduce urinary 8-OH-dG levels, a measure of oxidative damage. A phase I study was performed to determine the maximum tolerated dose of green tea extracts that contained 13.9% EGCG, 17.7% of other catechins, and 6.8% caffeine.[66] Sixteen patients who were diagnosed with lung cancer were administered the green tea extracts using an accelerated dosing regimen starting with 0.5 g/m^2/day. Adverse effects experienced by the patients included diarrhea, nausea, hypertension, heartburn and insomnia. The maximum tolerated dose was determined to be 3 g/m^2/day.

Chemoprevention using micronutrients. Chemoprevention trials designed to determine the effectiveness of micronutrients in preventing lung and prostate cancer have led to disappointing results.[38] The Alpha-Tocopherol, Beta-Carotene Cancer Prevention Study and Carotene and Retinol Efficacy Trial were based primarily on results obtained from epidemiological and laboratory-based studies that suggested a protective role of α-tocopherol (vitamin E) and β-carotene, a precursor of vitamin A. In the Alpha-Tocopherol, Beta-Carotene Cancer Prevention Study, a six-year study involving 29,133 Finnish male smokers, β-carotene was found to increase the incidence of lung cancer.[67] A similar increase in lung cancer incidence in smokers who received β-carotene supplements was observed in the Carotene and Retinol Efficacy Trial.[68] A follow-up study indicated that β-carotene supplementation also increased total mortality, but this was due primarily to cardiovascular diseases.[69] Post-trial analyses also indicated that supplementation with α-tocopherol may decrease prostate cancer mortality.[70] The prevailing current notion is that with respect to human consumption of micronutrients, a U-shaped dose response curve exists.[71] Thereby, nutritional supplementation is most beneficial to individuals who are under-nourished, but can be toxic to those who are over-nourished. Deciphering where, on this U-shaped curve, the nutrient levels of an individual should lie is a major challenge.

An additional large clinical trial investigated the impact of selenium and vitamin E supplementation on prostate cancer incidence.[72] This study involved 35,533 men between the ages of 50 and 55 years with no evidence of prostate cancer. Either selenium (200 μg/day) and/or vitamin E (400 IU/day) were administered for approximately five years. No significant decreases in prostate cancer incidences were observed in any group treated with either selenium, vitamin E, or both as compared to the placebo control. However, a non-significant (17%) increase in prostate cancer incidence was detected in the subjects receiving only vitamin E as compared to the placebo control. Additional analyses, which included quantification of toenail selenium levels revealed that selenium supplementation increased the risk of high grade prostate cancer in men with high selenium levels.[73] Further, supplementation with vitamin E enhanced the risk of developing prostate cancer in men with low selenium levels. Finally, it was reported that in men with non-metastatic prostate cancer, selenium supplementation at levels of 140 μg/day or more may increase prostate cancer mortality.[74] Taken together, the results obtained from chemoprevention trials using micronutrients indicate that their impact on human cancer incidences cannot be easily extrapolated from laboratory or epidemiological studies. In addition and contrary to popular belief, use of these "natural" agents have the potential for imposing harm.

Summary. As can be observed from the preceding sections and as expressed by previous authors,[38] the use of cancer chemopreventive agents has been fraught with both successes and failures. Careful examination of the studies discussed thus far provides a foundation with which to build future development of flavonoid-based chemopreventive therapies. Chemopreventive approaches that utilize naturally occurring dietary substances such as flavonoids have tremendous potential, as they can be cost effective and pose limited side effects. However, considerable uncertainties surround their bioavailability and effectiveness. The goal of the next chapters is to address this gap in our understanding by assessing our current state of knowledge and identifying our major challenges as well as pitfalls. Toward this end, we will begin by reviewing the basic mechanisms that underlie the carcinogenic process and the experimental evidence that supports the role of individual, isolated flavonoids in inhibiting each specific carcinogenic step. We will then examine the convergence of chronic inflammation and cancer and the considerable body of evidence supporting the role of flavonoids in suppressing inflammation. By next focusing on the impact of flavonoids on gastrointestinal and steroid-responsive cancers, we will be able to appreciate their underlying commonalities and differences and recognize the potential for using flavonoid-based therapies for inhibiting these cancers. Finally, at our journey's end, we will identify unanswered questions and challenges that must be addressed in order for flavonoid-based chemoprevention to become a reality.

References

1. Cheng JT. Review: Drug therapy in Chinese traditional medicine. *J Clin Pharmacol.* May 2000;40(5):445–450.
2. Parasuraman S, Thing GS, Dhanaraj SA. Polyherbal formulation: Concept of ayurveda. *Pharmacognosy Rev.* Jul 2014;8(16):73–80.
3. Mann RD. Whither therapeutics? An enquiry into drug use from historical principles: Part II. *J R Soc Med.* Jul 1986;79(7):418–422.
4. Bernhoft A. A brief review on bioactive compounds in plants. In: Bernhoft A, ed. *Bioactive Compounds in Plants-Benefits and Risks for Man and Animals* Oslow, Norway: The Norwegian Academy of Science and Letters; 2010: 11–17.
5. Wangensteen H, Klarpas L, Alamgir M, Samuelsen AB, Malterud KE. Can scientific evidence support using Bangladeshi traditional medicinal plants in the treatment of diarrhoea? A review on seven plants. *Nutrients* May; 5(5):1757–1800.
6. Amoa Onguene P, Ntie-Kang F, Lifongo LL, Ndom JC, Sippl W, Mbaze LM. The potential of anti-malarial compounds derived from African medicinal plants, Part I: A pharmacological evaluation of alkaloids and terpenoids. *Malar J.* 12:449.
7. Sokoloff B, Eddy WH, Redd JB. The biological activity of a flavonoid (vitamin "P") compound. *J Clin Invest.* Apr 1951;30(4):395–400.
8. Del Rio D, Rodriguez-Mateos A, Spencer JP, Tognolini M, Borges G, Crozier A. Dietary (poly)phenolics in human health: Structures, bioavailability, and evidence of protective effects against chronic diseases. *Antioxid Redox Signal.* May 10, 2013;18(14):1818–1892.
9. Kumar S, Pandey AK. Chemistry and biological activities of flavonoids: An overview. *Scientific World Journal* 2013:162750.
10. Stobiecki M, Kachlicki P. Isolation and Identification of Flavonoids. In: Grotewold E, ed. *The Science of Flavonoids.* New York, New York: Springer; 2006:47–69.
11. Maras JE, Talegawkar SA, Qiao N, Lyle B, Ferrucci L, Tucker KL. Flavonoid intakes in the Baltimore Longitudinal Study of Aging. *J Food Compost Anal.* Dec 1, 2011;24(8):1103–1109.
12. Beking K, Vieira A. An assessment of dietary flavonoid intake in the UK and Ireland. *Int J Food Sci Nutr.* Feb 2011;62(1):17–19.

13. Chun OK, Chung SJ, Song WO. Estimated dietary flavonoid intake and major food sources of U.S. adults. *The Journal of Nutrition.* May 2007;137(5):1244–1252.

14. Hollman PC, Katan MB. Dietary flavonoids: Intake, health effects and bio-availability. *Food Chem Toxicol.* Sep–Oct 1999;37(9–10):937–942.

15. Zamora-Ros R, Knaze V, Lujan-Barroso L, *et al.* Estimated dietary intakes of flavonols, flavanones and flavones in the European Prospective Investigation into Cancer and Nutrition (EPIC) 24 hour dietary recall cohort. *Br J Nutr.* Dec 2011;106(12):1915–1925.

16. Harnly JM, Doherty RF, Beecher GR, *et al.* Flavonoid content of U.S. fruits, vegetables, and nuts. *J Agric Food Chem.* Dec 27 2006;54(26):9966–9977.

17. Rasul A, Millimouno FM, Ali Eltayb W, Ali M, Li J, Li X. Pinocembrin: A novel natural compound with versatile pharmacological and biological activities. *BioMed Research International.* 2013;2013:379850.

18. Freischmidt A, Jurgenliemk G, Kraus B, *et al.* Contribution of flavonoids and catechol to the reduction of ICAM-1 expression in endothelial cells by a standardised Willow bark extract. *Phytomedicine: International Journal of Phytotherapy and Phytopharmacology.* Feb 15, 2012;19(3–4):245–252.

19. Mikulic-Petkovsek M, Slatnar A, Stampar F, Veberic R. HPLC-MSn identification and quantification of flavonol glycosides in 28 wild and cultivated berry species. *Food Chemistry.* Dec 15, 2012;135(4):2138–2146.

20. Lee J, Mitchell AE. Pharmacokinetics of quercetin absorption from apples and onions in healthy humans. *J Agric Food Chem.* Apr 18, 2012;60(15):3874–3881.

21. Slimestad R, Fossen T, Verheul MJ. The flavonoids of tomatoes. *J Agric. Food Chem.* Apr 9, 2008;56(7):2436–2441.

22. Biesaga M, Ochnik U, Pyrzynska K. Fast analysis of prominent flavonoids in tomato using a monolithic column and isocratic HPLC. *Journal of Separation Science.* Aug 2009;32(15–16):2835–2840.

23. Miean KH, Mohamed S. Flavonoid (myricetin, quercetin, kaempferol, luteolin, and apigenin) content of edible tropical plants. *J. Agric. Food Chem.* Jun 2001;49(6):3106–3112.

24. Rodriguez Galdon B, Rodriguez Rodriguez EM, Diaz Romero C. Flavonoids in onion cultivars (Allium cepa L.). *J. Food Sci.* Oct 2008;73(8):C599–605.

25. Yang RY, Lin S, Kuo G. Content and distribution of flavonoids among 91 edible plant species. *Asia Pacific Journal of Clinical Nutrition.* 2008; 17 Suppl 1:275–279.

26. Chirumbolo S. Anticancer properties of the flavone wogonin. *Toxicology.* Dec 6, 2013;314(1):60–64.

27. Nogata Y, Sakamoto K, Shiratsuchi H, Ishii T, Yano M, Ohta H. Flavonoid composition of fruit tissues of citrus species. *Bioscience, Biotechnology, and Biochemistry.* Jan 2006;70(1):178–192.

28. Han D, Row KH. Determination of luteolin and apigenin in celery using ultrasonic-assisted extraction based on aqueous solution of ionic liquid coupled with HPLC quantification. *Journal of the Science of Food and Agriculture.* Dec 2011;91(15):2888–2892.

29. Tomaino A, Martorana M, Arcoraci T, Monteleone D, Giovinazzo C, Saija A. Antioxidant activity and phenolic profile of pistachio (Pistacia vera L., variety Bronte) seeds and skins. *Biochimie.* Sep 2010;92(9):1115–1122.

30. Costa P, Goncalves S, Valentao P, *et al.* Metabolic profile and biological activities of Lavandula pedunculata subsp. lusitanica (Chaytor) Franco: studies on the essential oil and polar extracts. *Food Chemistry.* Dec 1, 2013;141(3):2501–2506.

31. Plazonic A, Bucar F, Males Z, Mornar A, Nigovic B, Kujundzic N. Identification and quantification of flavonoids and phenolic acids in burr parsley (Caucalis platycarpos L.), using high-performance liquid chromatography with diode array detection and electrospray ionization mass spectrometry. *Molecules.* 2009;14(7):2466–2490.

32. Guimaraes R, Barros L, Duenas M, *et al.* Nutrients, phytochemicals and bioactivity of wild Roman chamomile: a comparison between the herb and its preparations. *Food Chemistry.* Jan 15 2013;136(2):718–725.

33. Mortensen A, Kulling SE, Schwartz H, *et al.* Analytical and compositional aspects of isoflavones in food and their biological effects. *Mol Nutr Food Res.* Sep 2009;53 Suppl 2:S266–309.

34. Vogiatzoglou A, Mulligan AA, Luben RN, *et al.* Assessment of the dietary intake of total flavan-3-ols, monomeric flavan-3-ols, proanthocyanidins and theaflavins in the European Union. *The British Journal of Nutrition.* Apr 28, 2014;111(8):1463–1473.

35. Zamora-Ros R, Knaze V, Lujan-Barroso L, *et al.* Estimation of the intake of anthocyanidins and their food sources in the European Prospective Investigation into Cancer and Nutrition (EPIC) study. *The British Journal of Nutrition.* Oct 2011;106(7):1090–1099.

36. Park EJ, Pezzuto JM. Botanicals in cancer chemoprevention. *Cancer Metastasis Reviews.* 2002;21(3–4):231–255.

37. Lippman SM, Hawk ET. Cancer prevention: from 1727 to milestones of the past 100 years. *Cancer Res.* Jul 1, 2009;69(13):5269–5284.

38. Patterson SL, Colbert Maresso K, Hawk E. Cancer chemoprevention: Successes and failures. *Clinical Chemistry.* Jan 2013;59(1):94–101.

39. Giardiello FM, Hamilton SR, Krush AJ, *et al.* Treatment of colonic and rectal adenomas with sulindac in familial adenomatous polyposis. *The New England Journal of Medicine.* May 6, 1993;328(18):1313–1316.

40. Greenberg AK, Tsay JC, Tchou-Wong KM, Jorgensen A, Rom WN. Chemoprevention of lung cancer: Prospects and disappointments in human clinical trials. *Cancers.* 2013;5(1):131–148.

41. Jordan VC. Tamoxifen: A most unlikely pioneering medicine. *Nature Reviews. Drug Discovery.* Mar 2003;2(3):205–213.

42. Siegfried JM. Smoking out reproductive hormone actions in lung cancer. *Molecular Cancer Research: MCR.* Jan 2014;12(1):24–31.

43. Caiazza F, Ryan EJ, Doherty G, Winter DC, Sheahan K. Estrogen receptors and their implications in colorectal carcinogenesis. *Frontiers in Oncology.* 2015;5:19.

44. Latte KP, Appel KE, Lampen A. Health benefits and possible risks of broccoli — an overview. *Food Chem Toxicol.* Dec 2011;49(12):3287–3309.

45. Hecht SS. Inhibition of carcinogenesis by isothiocyanates. *Drug Metabolism Reviews.* Aug–Nov 2000;32(3–4):395–411.

46. Zhang Y, Kensler TW, Cho CG, Posner GH, Talalay P. Anticarcinogenic activities of sulforaphane and structurally related synthetic norbornyl isothiocyanates. *Proceedings of the National Academy of Sciences of the United States of America.* Apr 12, 1994;91(8):3147–3150.

47. Bhandari SR, Kwak JH. Chemical composition and antioxidant activity in different tissues of brassica vegetables. *Molecules.* 2015;20(1):1228–1243.

48. Cartea ME, Francisco M, Soengas P, Velasco P. Phenolic compounds in Brassica vegetables. *Molecules.* 2011;16(1):251–280.

49. Higdon JV, Delage B, Williams DE, Dashwood RH. Cruciferous vegetables and human cancer risk: Epidemiologic evidence and mechanistic basis. *Pharmacological Research : The Official Journal of the Italian Pharmacological Society.* Mar 2007;55(3):224–236.

50. Shimamoto K, Dewa Y, Ishii Y, *et al*. Indole-3-carbinol enhances oxidative stress responses resulting in the induction of preneoplastic liver cell lesions in partially hepatectomized rats initiated with diethylnitrosamine. *Toxicology*. May 10 2011;283(2–3):109–117.

51. Shilling AD, Carlson DB, Katchamart S, Williams DE. 3,3′-diindolylmethane, a major condensation product of indole-3-carbinol, is a potent estrogen in the rainbow trout. *Toxicol Appl Pharmacol*. Feb 1, 2001;170(3):191–200.

52. Fujioka N, Ainslie-Waldman CE, Upadhyaya P, *et al*. Urinary 3,3′-diindolylmethane: A biomarker of glucobrassicin exposure and indole-3-carbinol uptake in humans. *Cancer epidemiology, biomarkers & prevention: A publication of the American Association for Cancer Research, cosponsored by the American Society of Preventive Oncology*. Feb 2014;23(2):282–287.

53. Kensler TW, Chen JG, Egner PA, *et al*. Effects of glucosinolate-rich broccoli sprouts on urinary levels of aflatoxin-DNA adducts and phenanthrene tetraols in a randomized clinical trial in He Zuo township, Qidong, People's Republic of China. *Cancer epidemiology, biomarkers & prevention: A publication of the American Association for Cancer Research, cosponsored by the American Society of Preventive Oncology*. Nov 2005;14(11 Pt 1):2605–2613.

54. Kensler TW, Ng D, Carmella SG, *et al*. Modulation of the metabolism of airborne pollutants by glucoraphanin-rich and sulforaphane-rich broccoli sprout beverages in Qidong, China. *Carcinogenesis*. Jan 2012;33(1):101–107.

55. Stoner GD, Wang LS, Casto BC. Laboratory and clinical studies of cancer chemoprevention by antioxidants in berries. *Carcinogenesis*. Sep 2008;29(9):1665–1674.

56. Stoner GD, Sardo C, Apseloff G, *et al*. Pharmacokinetics of anthocyanins and ellagic acid in healthy volunteers fed freeze-dried black raspberries daily for 7 days. *Journal of Clinical Pharmacology*. Oct 2005;45(10):1153–1164.

57. Mallery SR, Zwick JC, Pei P, *et al*. Topical application of a bioadhesive black raspberry gel modulates gene expression and reduces cyclooxygenase 2 protein in human premalignant oral lesions. *Cancer Res*. Jun 15, 2008;68(12):4945–4957.

58. Lambert JD. Does tea prevent cancer? Evidence from laboratory and human intervention studies. *The American Journal of Clinical Nutrition*. Dec 2013;98(6 Suppl):1667S–1675S.

59. Sang S, Lambert JD, Ho CT, Yang CS. The chemistry and biotransformation of tea constituents. *Pharmacological research: The Official Journal of The Italian Pharmacological Society*. Aug 2011;64(2):87–99.

60. Moses MA, Henry EC, Ricke WA, Gasiewicz TA. The heat shock protein 90 inhibitor, (–)-epigallocatechin gallate, has anticancer activity in a novel human prostate cancer progression model. *Cancer Prev Res (Phila).* Mar 2015;8(3):249–257.

61. Hara Y. Tea catechins and their applications as supplements and pharmaceutics. *Pharmacological Research: The Official Journal of The Italian Pharmacological Society.* Aug 2011;64(2):100–104.

62. Garcia FA, Cornelison T, Nuno T, *et al.* Results of a phase II randomized, double-blind, placebo-controlled trial of Polyphenon E in women with persistent high-risk HPV infection and low-grade cervical intraepithelial neoplasia. *Gynecologic Oncology.* Feb 2014;132(2):377–382.

63. Yoon AJ, Shen J, Santella RM, *et al.* Topical Application of Green Tea Polyphenol (–)-Epigallocatechin-3-gallate (EGCG) for Prevention of Recurrent Oral Neoplastic Lesions. *Journal of Orofacial Sciences.* 2012;4(10):43–50.

64. Rhodes LE, Darby G, Massey KA, *et al.* Oral green tea catechin metabolites are incorporated into human skin and protect against UV radiation-induced cutaneous inflammation in association with reduced production of pro-inflammatory eicosanoid 12-hydroxyeicosatetraenoic acid. *The British Journal of Nutrition.* Sep 14 2013;110(5):891–900.

65. Hakim IA, Harris RB, Brown S, *et al.* Effect of increased tea consumption on oxidative DNA damage among smokers: A randomized controlled study. *The Journal of Nutrition.* Oct 2003;133(10):3303S–3309S.

66. Laurie SA, Miller VA, Grant SC, Kris MG, Ng KK. Phase I study of green tea extract in patients with advanced lung cancer. *Cancer Chemotherapy and Pharmacology.* Jan 2005;55(1):33–38.

67. Alpha-Tocopherol BCCPSG. The effect of vitamin E and beta carotene on the incidence of lung cancer and other cancers in male smokers. The Alpha-Tocopherol, Beta Carotene Cancer Prevention Study Group. *The New England Journal of Medicine.* Apr 14 1994;330(15):1029–1035.

68. Omenn GS, Goodman GE, Thornquist MD, *et al.* Risk factors for lung cancer and for intervention effects in CARET, the Beta-Carotene and Retinol Efficacy Trial. *Journal of the National Cancer Institute.* Nov 6, 1996;88(21):1550–1559.

69. Virtamo J, Pietinen P, Huttunen JK, *et al.* Incidence of cancer and mortality following alpha-tocopherol and beta-carotene supplementation: A postintervention follow-up. *Jama.* Jul 23, 2003;290(4):476–485.

70. Virtamo J, Taylor PR, Kontto J, *et al*. Effects of alpha-tocopherol and beta-carotene supplementation on cancer incidence and mortality: 18-year postintervention follow-up of the Alpha-tocopherol, Beta-carotene Cancer Prevention Study. *Int J Cancer.* Jul 1, 2014;135(1):178–185.

71. Brasky TM, Kristal AR. Learning from history in micronutrient research. *Journal of the National Cancer Institute.* Jan 2015;107(1):375.

72. Lippman SM, Klein EA, Goodman PJ, *et al*. Effect of selenium and vitamin E on risk of prostate cancer and other cancers: The Selenium and Vitamin E Cancer Prevention Trial (SELECT). *Jama.* Jan 7, 2009;301(1):39–51.

73. Kristal AR, Darke AK, Morris JS, *et al*. Baseline selenium status and effects of selenium and vitamin E supplementation on prostate cancer risk. *Journal of the National Cancer Institute.* Mar 2014;106(3):djt456.

74. Kenfield SA, Van Blarigan EL, DuPre N, Stampfer MJ, E LG, Chan JM. Selenium supplementation and prostate cancer mortality. *Journal of the National Cancer Institute.* Jan 2015;107(1):360.

Mechanisms by Which Flavonoids Exert their Beneficial Anti-cancer Effects*

2

The hallmarks of cancer. The development of the fully malignant disease state of cancer as described in an authoritative review by Hanahan and Weinberg encompasses six biological processes termed "hallmarks".[1] Here, a hallmark is a trait that governs the transformation of normal cells into cancerous ones. The six hallmarks of cancer are (1) sustained proliferative signaling, (2) evasion of growth suppression, (3) resistance to cell death, (4) attainment of replicative immortality, (5) attainment of angiogenesis, and (6) activation of invasion and metastasis. Two "emerging" hallmarks and two "enabling" characteristics are also important. An

---

*__Abbreviations__: AHR, aryl hydrocarbon receptor; AKT, v-akt murine thymoma viral oncogene homolog 1; ARNT, aryl hydrocarbon receptor translocator; BAK1, BCL2-antagonist/killer 1; BAX, BCL2-associated X protein; BCL2, B-cell lymphoma 2; CAR, constitutive androstane receptor; CYP1A1, cytochrome P4501A1; CYP1B1, cytochrome P4501B1; EGFR, epidermal growth factor receptor; ERBB2, erb-b2 receptor tyrosine kinase 2; ERK, extracellular signal-regulated kinase; FabG, 3-oxoacyl-[acyl-carrier-protein] reductase; FabI, enoyl-[acyl-carrier-protein] reductase; FGF, fibroblast growth factor; HUVEC, human umbilical vein endothelial cells; IGF1R, insulin-like growth factor 1; IL6, interleukin 6; KEAP1, Kelch-like ECH-associated protein 1; MAPK, mitogen activated protein kinases; MAPK8, mitogen-activated protein kinase 8; PIK3CA, phosphatidylinositol-4,5-bisphosphate 3-kinase, catalytic subunit alpha; mTOR, mechanistic target of rapamycin; NFκB, nuclear factor of kappa light polypeptide gene enhancer in B-cells 1; factor of kappa light polypeptide gene enhancer in B-cells; NRF2, nuclear factor (erythroid-derived 2)-like 2; PUMA, TP53-upregulated modulator of apoptosis; PXR, pregnane X receptor; RAS, Rat sarcoma; RB, retinoblastoma; RXR, retinoid X receptor; ROS, reactive oxygen species; STAT3, signal transducer and activator of transcription 3; TGFβ, transforming growth factor beta; TP53, tumor protein 53; TRAMP, transgenic adenocarcinoma of the mouse prostate; TSP-1, thrombospondin-1; VEGF, vascular endothelial growth factor.

"emerging" hallmark is a functionally important attribute of a neoplastic cell. The two that arise most frequently are **deregulated cellular energetics** and **avoidance of immune destruction**. Finally, "enabling" characteristics are those processes which make the attainment of the six hallmarks possible. These are **genomic instability and mutations** and **tumor-promoting inflammation**. As we consider the putative mechanisms by which flavonoids are thought to inhibit and/or prevent the formation and progression of cancer, we will use the framework provided by these described hallmarks as our guide.

Impact of flavonoids on sustained proliferative signaling. Sustained proliferative signaling occurs most frequently via excessive production of growth factors, ligand-independent activation of growth factor receptors, and receptor-independent activation of downstream signaling. Dysregulated activation of the epidermal growth factor receptor (EGFR) and its close relatives, such as ERBB2 (also known as HER2, human epidermal growth factor 2) occurs in a wide variety of epithelial-derived tumors, in particular that of the lung (EGFR), colon (EGFR) and breast (ERBB2).[2] With respect to growth factor receptor activation, a number of flavonoids have been shown to suppress activation of the EGFR including fisetin, naringenin, apigenin, quercetin, verbacoside, rutin, luteolin and xanthohumol.[3–10] Finally, quercetin and apigenin have been shown to inhibit ERBB2 and IGF1R, respectively.[11,12]

The uncoupling of the growth factor receptors from their downstream signaling pathways is a frequent event that occurs during the tumorigenic process. Typically, receptor-independent activation of signaling pathways such as the MAPK and PIK3CA/AKT/mTOR signaling pathways occurs.[1] In this manner, the negative feedback systems designed to protect against excessive receptor signaling are disrupted. The MAPK pathway is composed of three subfamilies: the ERK (extracellular-signal regulated kinase, also referred to as the RAS-RAF-MEK-ERK pathway), MAPK8 (also known as JNK) and p38 MAPK.[13,14] The ERKs are perhaps best characterized as key responders to mitogenic stimuli. They are the downstream signaling components that mediate the pro-tumorigenic effects of oncogenic RAS, a major driver of tumor formation. The three RAS genes, *HRAS*, *NRAS,* and *KRAS* are the most frequently mutated oncogenes, occurring in approximately 30% of all human tumors.[15] The ultimate effect that arises following activation of the ERK pathway is context-dependent where the signal intensity is thought to determine whether a pro-proliferative or growth suppressive response ensues. A number of cancer cells harbor constitutive activation of the ERK pathway which is linked to their survival. In contrast to ERK, the MAPK8 and p38 MAPK proteins are highly activated by stress signaling and inflammatory cytokines. Upregulation of this pathway often leads to the induction of apoptosis rather than proliferation.

Activation of G protein coupled receptors and receptor tyrosine kinases by growth factors results in the upregulation of the PIK3CA/AKT/mTOR pathway (also referred to as the PI3K pathway).[16] Like the ERK pathway, the PIK3CA pathway is often constitutively active in a number of human cancers and is associated with tumor promotion and maintenance of the tumor phenotype. Further, it is a key regulator of cell growth, survival, and proliferation. The ERK and PIK3CA pathways converge on mTOR which is the master regulator of protein translation. Thus, considerable crosstalk exists between these pathways in their regulation of cell fate decisions.

The impact of flavonoids on ERK signaling includes both activation as well as suppression of the signaling pathway. For example, luteolin 8-C-β fucopyranoside, kaempferol, isoorientin, baicalein, quercetin, and myricetin have all been shown to suppress phosphorylation of components of the ERK pathway.[17-24] Conversely, other studies have reported that flavonoids such as genistein, diosmetin, luteolin, and procyanidin B2 upregulate ERK signaling.[25-27] The impact of apigenin appears to be somewhat unique as it has been described as causing an "unbalanced" activation of ERK1/2.[28] That is, it causes sustained phosphorylation that is incapable of impacting downstream phosphorylation events. With respect to p38 MAPK signaling, fisetin and quercetin suppress p38 MAPK phosphorylation,[22,29] whereas isoorientin, procyanidin B2, quercetin, and trifolin enhance its phosphorylation.[27,30-32] Interestingly, the ability of quercetin to induce p38 MAPK kinase phosphorylation appears to be dependent on whether or not the cells express estrogen receptor α.[33] Similar conflicting effects of flavonoids have also been observed with respect to MAPK8 signaling where quercetin, trifolin, and isoorientin have been reported to induce MAPK8 phosphorylation[19,31,32] while acacetin and quercetin inhibit[22,34] MAPK8 phosphorylation. Flavonoids like quercetagetin appear to inhibit MAPK8 via direct binding to its ATP-binding site.[35] A wide variety of flavonoids representing the majority of the flavonoid subclasses have been shown to inhibit the PIK3CA/AKT/mTOR pathway as follows: flavonols (i.e., dihydromyricetin, myricetin, fisetin, isorhamnetin, kaempferol, quercetin, and taxifolin), flavones (i.e., apigenin, baicallin, chrysin, luteolin, nobilein, tangeretin, and wogonin), isoflavones (i.e., formononetin, pomiferin triacetate, genistein and 7,3′,4′-trihydroxyisoflavone, a metabolite of daidzein), flavan-3-ols (epigallocatechin and -epigallocatechin-3-gallate), and anthocyanidins (delphinidin).[12,36–55] In addition, members of the deoxychalcones (i.e., butein and isoliquiritigenin), which are structurally similar to the flavonoids but lack the C ring, have also been shown to inhibit the PIK3CA/AKT/mTOR pathway.[44,48]

Impact of flavonoids on evasion of growth suppression. A tumor cell acquires its ability to evade growth suppressors in part, by inactivating genes identified as tumor suppressors, such as RB1 (retinoblastoma-associated) and TP53.[1] The proteins encoded by these genes play central roles in governing cell fate decisions. Other growth suppressors whose activities become abrogated during tumorigenesis include the TGFβ signaling pathway and gene products involved in maintaining contact inhibition. A number of flavonoids have been shown to activate the TP53 pathway by increasing its expression and/or increasing its phosphorylation status.[42,56–64] Included in this group are dihydromyricetin, fisetin, apigenin, quercetin, wogonin, naringenin, luteolin, epicatechin gallate, neoheperidin, celastrol, and vitexin. A limited number of flavonoids have also been found to decrease the expression and/or phosphorylation of RB and include baicalein, quercetin, luteolin, isoliquiritigenin, and apigenin.[65–69] While detailed mechanistic information is yet to be determined, it appears that the impact of flavonoids on TP53 and RB signaling is most likely indirect and involves perturbation of the above mentioned kinase pathways. In some cases (i.e., baicalein-initiated events), the aryl hydrocarbon receptor (a ligand-activated transcription factor to be discussed below) has been implicated.[65]

Impact of flavonoids on resistance to cell death. Programmed cell death by apoptosis forms an important barrier to cancer development.[1] Thus, overcoming this natural barrier is often a strategy that is used by a successful tumor cell. While two major pathways mediate apoptosis, the intrinsic and extrinsic pathways, it is the intrinsic apoptotic program that is thought to be of primary importance for the developing tumor. The major regulatory proteins that comprise the pro- versus anti-apoptotic forces include BCL2 ("anti") and BAX, BAK1, and cytochrome c ("pro"). In addition, TP53, in its role as a DNA damage sensor protein, often triggers apoptosis via upregulation of proteins such as PUMA. A second strategy used by the tumor cell to maintain its survival is to induce autophagy, which like apoptosis, is often induced by cellular stress. In some cases, such as the presence of cytotoxic drugs, autophagy is closely linked to cell survival. Many of the signaling pathways that trigger apoptosis are also capable of inducing autophagy, depending on the context of the cellular environment.

A major mechanism by which flavonoids are thought to exert their chemopreventive effects is via induction of apoptosis in the tumor cell. Flavonoids with demonstrated apoptosis-inducing properties include acacetin,[70] alpinetin,[71] apigenin,[72] baicalein,[73] butein,[74] epicatechin gallate,[42] fisetin,[75] galangin,[76] herbacetin,[77] hesperetin,[78] hispidulin,[79] isoliquiritigenin,[80] isoorientin,[19] luteolin,[81] kaempferol,[82] kurarinol,[83] myricetin,[84] naringenin,[61] neohesperidin,[63] rutin,[85] quercetin,[59] tamarixetin,[86] wogonin,[60] and wogonoside.[87] In some cases (i.e., apigenin,[58] baicalein,[88] fisetin,[89] galangin,[90] quercetin,[59] luteolin,[81] wogonoside[87]), treatment of cultured cells with flavonoids can also induce autophagy. Often the concentrations used for flavonoid-induced apoptosis are quite high, at times exceeding 100 μM. Since these concentrations would be difficult to achieve *in vivo* within the tumor tissue, it is important to obtain *in vivo* evidence of flavonoid-induced apoptosis. Typically, an athymic mouse (that is immunodeficient) is orally administered the flavonoid of interest (i.e., acacetin,[70] ampelopsin,[91] apigenin,[92–94] kaempferol,[95] myricetin,[96] oroxylin A,[97] orquercetin[98]) and innoculated with a cultured human tumor cell line. The growth of this explanted tumor is then monitored and the rate of growth is calculated relative to the growth of tumors in mice that were not administered the flavonoid. At the end of the experiment, the tumors are removed and examined for changes in the expression levels of proteins known to be involved in apoptosis (i.e., TP53, BCL, BAX, or BAK). Thus, by using explanted cultured human tumor cells, it can be demonstrated that the flavonoid, at a given administered dose, is capable of reaching the tumor cell and inducing apoptosis like that observed in the tissue culture dish.

An additional approach that may be used for demonstrating that the pro-apoptotic actions of a particular flavonoid occur *in vivo* is to administer the flavonoid to a rat or mouse that has been exposed to a carcinogen to initiate formation of a mutated tumor cell. Typically, the flavonoid is administered throughout the course of the entire tumorigenic process. Using this type of approach, quercetin has been shown to inhibit hepato-carcinogenesis,[99] and baicalein has been shown to inhibit colorectal cancer[100] in a manner that was accompanied by increases in the expression levels of apoptotic markers. Use of this type of *in vivo* approach versus that using the explanted tumor cells allows for more mechanistic information regarding the chemopreventive actions of the flavonoids to be gleaned. However, a disadvantage is that these models may not accurately recapitulate the events that occur during human carcinogenesis and species-specific differences that exist may complicate data interpretation.

Impact of flavonoids on attainment of replicative immortality. As normal cells progressively divide, each generation typically harbors successively shorter telomeric DNA. When a critical length is ultimately reached, crises, followed by senescence and apoptosis ensue.[1] To overcome this series of events, a successful tumor cell frequently acquires high telomerase activity which allows the tumor cells to maintain telomeric DNA at lengths that are incapable of triggering senescence and/or apoptosis. A number of flavonoids target this survival mechanism by attenuating the tumor cell's relatively high levels of telomerase activity. Included here are apigenin,[101] butein,[102] epigallocatechin-3-gallate,[103] genistein,[104] quercetin,[105] and wogonin.[106] Interestingly, with respect to epigallocatechin-3-gallate, its ability to inhibit telomerase activity appears to be its primary mode of chemopreventive action.[107]

Impact of flavonoids on attainment of angiogenesis. A successful tumor cell ensures that it has a constant supply of nutrients that is adequate for maintaining its growth.[1] It accomplishes this feat by developing its own supply of capillaries and blood vessels. The vascular system within a tumor is poorly organized and is characterized by high vascular permeability and high resistance to blood flow.[108] In addition, the endothelial cells within tumors are highly prone to active proliferation and thereby actively participate in angiogenesis. Tumor angiogenesis most commonly involves upregulation of angiogenesis inducers — vascular endothelial growth factor (VEGF-A) and fibroblast growth factor (FGF) — as well as suppression of angiogenesis inhibitors (i.e., thrombospondin-1, TSP-1). However, the importance of the considerable interplay between the immune cells within the tumor microenvironment and the endothelial cells should also be realized.[109] The impact of flavonoids on angiogenesis has been examined by observing their effects on the growth and invasion of cultured cells, in particular human umbilical vein endothelial cells (HUVEC). Other approaches include using an aortic ring model to evaluate the capacity of either rat or mouse aortic explants to form new vessels in gels of collagen, fibrin, or basement membranes.[110] The aortic ring model is thought to represent a combination of the advantages posed by both *in vitro* and *in vivo* approaches. Angiogenesis that arises from the new blood vessels formed within human tumors implanted either subcutaneously or within the tissue of interest in a nude mouse or within an animal model of tumorigenesis also provides a reasonable model of tumor-driven angiogenesis. Given the impact of the host environment on tumor angiogenesis, the site of tumor implantation can have a significant impact on the events associated with tumor vascularization and should be recognized when interpreting the results obtained from these different types of studies.[108] Finally, an additional *in vivo* model of angiogenesis that is rapidly gaining acceptance within the scientific community is the use of zebrafish.[111] Zebrafish offer a number of advantages as an experimental *in vivo* model, in particular, when the developing embryo is used. It is low-cost and allows for rapid screening of drugs and phytochemicals in a fully functioning, interconnected organ system. Similar advantages may also be gained when using the chick embryo as an experimental model of angiogenesis.[112]

Using cultured HUVEC cells, flavonoids such as baicalein,[113] glabridin,[114] hispidulin,[115] 4-hydrozychalcone,[116] isoliquiritigenin,[117] nobiletin,[118] quercetin,[119] and wogonin[120] have been shown to inhibit many *in vitro* events associated with angiogenesis (i.e., migration, tube formation, etc.) and block VEGF signaling. Flavonoids with anti-angiogenic properties demonstrated using the rat aortic ring model include baicalein,[113] hispidulin,[115] isoliquiritigenin,[44] and luteolin.[121] Flavonoids exerting anti-angiogenic effects in the zebrafish model include nobiletin[118] and quercetin[119] and in the chick embryo model include apigenin,[122] kaempferol,[123] luteolin,[121] 4-hydoxychalcone,[116] and wogonin.[120] In a screen of seven polymethoxylated flavonoids including hesperetin, naringin, nobiletin, and sinensetin that employed both the HUVEC and zebrafish models of angiogenesis, sinensetin was identified as the most potent anti-angiogenic flavonoid.[124] Finally, using explanted human tumor cells, the anti-angiogenic properties of flavonoids such as luteolin[121] and ampelopsin[91] (prostate), hispidulin[115] (pancreatic), glabridin[114] (breast) and apigenin (lung)[125] have been demonstrated. In the majority of these studies, flavonoid-inhibition of VEGF signaling has been the proposed underlying mechanism, but flavonoid-induced interference of the IL6/STAT3 pro-inflammatory pathway has also been shown.[126]

Impact of flavonoids on activation of invasion and metastasis. Metastasis, the final state of cancer progression, is composed of five discrete steps: loss of cellular adhesion, increased motility and invasiveness, entry and survival in the circulation, exit into new tissue and finally, colonization of a distant site.[127] It is also becoming increasingly clear that cancer is a systemic disease in which the tumor-host interactions and tumor-derived factors extend beyond the boundaries of individual cells to impact distant tissues.[128] Given the multi-step and systemic nature of tumor invasion and metastasis, we will not consider the plethora of studies that have been performed using cultured cells to examine the impact of flavonoids on isolated events related to the metastatic process. Instead, we will focus our attention on studies that have relied on *in vivo* approaches. *In vivo* tumor metastasis is often studied by injecting human tumors into either the heart or the tail vein of an athymic mouse and observing the dissemination of the tumor cells into various tissues.

Alternatively, tumor cells that spontaneously disseminate from a tumor explanted either subcutaneously or within specific tissue sites can be monitored. The dissemination of the tumor cells from an originating tumor to other tissues within genetically engineered models of human cancer, such as the TRAMP model (discussed in Chapter 5) of prostate cancer, has also been used to study tumor metastasis. Flavonoids that have been shown to inhibit metastasis using these *in vivo* models include ampelopsin,[91] apigenin,[129,130] chrysin,[131] deguelin,[132] galangin,[133] isoliquiritigenin,[134] luteolin,[135] luteoloside,[136] naringenin,[137] and silibinin.[138] The proposed mechanisms by which flavonoids are thought to inhibit metastasis include modulation of the activity and/or expression levels of cell adhesion molecules (i.e., β-catenin, metallothiein matrix proteinases, and focal adhesion kinase), inhibition of VEGF-induced angiogenesis, and decreased immunosuppression.

Summary of the impact of flavonoids on the hallmarks of cancer. The discussion of the evidence generated thus far indicates that flavonoids are capable of impinging on all of the hallmarks of cancer. This is depicted in Figure 2.1. However, it is also possible that inflammation plays an underlying role in cancer progression and that flavonoids exert their primary effects via their ability to inhibit inflammation. The evidence supporting this idea is discussed in Chapter 3. Finally, the impact of flavonoids on other mechanisms that can contribute to the carcinogenic process is discussed below.

Additional cancer-relevant mechanisms. Mechanisms in addition to those discussed above that are thought to underlie the chemopreventive effects of flavonoids include their ability to abrogate events associated with oxidative stress and alter pathways associated with the metabolism and excretion of drugs and xenobiotics. The importance of these flavonoid-induced events relative to those discussed above is currently under extensive research and discussion. However, many of these events are intertwined and are representative of a series of protective events that coordinately inhibit the adverse effects initiated by oxidative stress and inflammation. Of particular interest is the impact of flavonoids on the NRF2, PXR, CAR, AHR, and NFκB pathways, as well as the nuclear

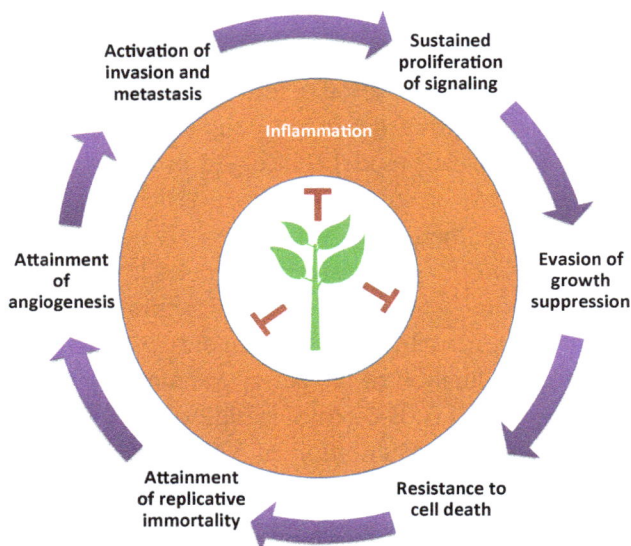

Fig. 2.1. Proposed mechanisms by which flavonoids inhibit carcinogenesis.
Flavonoids may protect against cancer initiation and development by inhibiting all of the
events described as the "hallmarks of cancer"[1] as well as inhibiting inflammation.

steroid receptors. The NFκB pathway and its putative role in mediating
the anti-inflammatory effects of flavonoids will be addressed in Chapter 3,
whereas flavonoid modulation of steroid receptors will be discussed in
Chapter 5.

Flavonoids and oxidative stress. The idea that the formation of free radicals and reactive oxygen species (ROS) play a key role in normal biological processes as well as the development of pathological conditions
associated with aging and diseases gained traction in the 1950s.[139–141]
Initially, it was thought that free radicals and ROS played only deleterious
roles. However, it is now known that the generation of ROS is required for
normal signal propagation and for maintaining tissue homeostasis.
Dysregulation of tissue homeostasis often results in the generation of ROS
that exceeds the capacity of antioxidant defenses. Ultimately, the oxidative stress that is formed damages macromolecules such as DNA, proteins,
and lipids. ROS is produced by the mitochondria via electrons that escape

the electron transport chain, by oxidases and cytochrome P450 enzymes, and following the auto-oxidation of xenobiotics and endogenous substances such as epinephrine. Immune cells such as neutrophils also generate large amounts of ROS as they undergo phagocytosis of microorganisms and cellular debris. Increased cellular levels of ROS facilitate the downstream events that occur following growth factor receptor activation and propagate the signal transduction cascade. Immune cells such as macrophages and T cells rely on ROS for their differentiation and maturation. Commensal bacteria stimulate epithelial ROS production to promote intestinal stem cell proliferation and thereby contribute to the maintenance of gut homeostasis. This likely involves an interplay between signaling molecules such as NFκB, β-catenin, and nuclear receptors such as the PXR and perhaps the AHR.

The antioxidant defenses are composed of molecules with low and high molecular mass. Low molecular mass antioxidants include substances such as ascorbic acid, tocopherol, and glutathione. High molecular mass antioxidants are enzymes such as superoxide dismutase and glutathione reductase. Excessive amounts of ROS enhance the expression levels of antioxidant enzymes which during normal conditions enhance the cell's capacity for eliminating excess ROS. Under conditions of chronic oxidative stress, which often occurs during chronic infections and within chronically inflamed tissues, the enhanced levels of antioxidant systems represent an adaptation to the higher levels of ROS. The NRF2, NFκB, AP-1 and MAPK signaling systems facilitate cellular adaption to high ROS. It is thought that low and acute levels of ROS are beneficial as they develop tolerance by expanding the cell's antioxidant defense capacity. However, the high and sustained levels of ROS that exceed this capacity have pathological consequences.

Despite the evidence linking chronic ROS to the development of pathological disease states such as cancer, the use of antioxidant supplements designed to prevent cancer in clinical trials have either failed to demonstrate a beneficial effect or have resulted in enhanced cancer risk (nccam. nih.gov, accessed January 2015). It may be that the effectiveness of antioxidant supplements is dependent on the stage of the carcinogenic process. That is, their use may be effective only at the early stage while the tumor cell has yet to develop sufficient adaptive barriers to high ROS levels or when ROS-induced cellular damage is limited. Alternatively, the presence of the antioxidant within the tumor cells may disrupt the ROS-TP53 axis, thereby allowing the tumor cells to proliferate.[142]

While a considerable body of research has reported on the *in vitro* antioxidant properties of flavonoids, a recent review suggests that this mechanism may not play a major role in flavonoid chemoprevention.[143] The rationale is that dietary consumption of flavonoids results in relatively low (<1 μM) tissue concentrations and are typically in the presence of more effective antioxidants, such as glutathione. Thus, these relatively low levels of less effective antioxidants would be poor competitors with the actions of the more potent and abundant antioxidant, glutathione. Given this, it is unlikely that flavonoids act as effective free radical scavengers *in vivo*. A more plausible scenario is that the antioxidant and chemopreventive properties of flavonoids arise from their effects on signaling transduction systems as discussed above, as well as their impact on nuclear receptors and other transcription factors. An alternative explanation is that the antioxidant activities of flavonoids become important only when their tissue levels are relatively high and when the levels of the endogenous antioxidants like glutathione are compromised.

Flavonoids as antibacterial agents. Plants have developed a unique advantage in warding off pathogenic assaults via their production of flavonoids as secondary metabolites.[144] The first and oldest flavonoids synthesized by plants are chalcones, flavonols, and flavones. Flavone and a derivative of kaempferol (kaempferol-3-O-α-L (2′,4′-di-E-p-coumaroyl)-rhamnoside) are amongst the top ten most potent, naturally occurring flavonoids with respect to their anti-bacterial activities. The specific mechanisms by which flavonoids exert their anti-bacterial activities are yet to be fully elucidated.[145] Thus far, multiple mechanisms of action are thought to be involved and include (1) damaging bacterial membranes by generating hydrogen peroxide, (2) inhibiting bacterial virulence factors, and (3) inhibiting key bacterial enzymes. By inhibiting these enzymes, flavonoids inhibit DNA synthesis (via inhibition of topoisomerase and/or dihydrofolate reductase), energy metabolism (via inhibition of ATP synthase), cell wall synthesis (via inhibition of D-alanine–D-alanine ligase) and cell membrane synthesis (via inhibition of FabG, FabI, etc.). However, since flavonoids have also been found to induce bacterial aggregation, it is not yet clear whether their ability to inhibit enzyme activity is due to a direct flavonoid–enzyme interaction or occurs as a mechanism that is secondary to the aggregation effect.

Impact of drug metabolizing enzymes and transporters on flavonoids. Like the antioxidant systems described above, drug metabolizing enzymes and transporters can be thought of as components of a protective and defensive response.[146] They are designed to enhance the elimination of foreign or xenobiotic substances. In addition to xenobiotics, drug metabolizing enzymes and transporters aid in the elimination of endogenous substances such as hormones and other lipophilic substances as well as pharmaceutical drugs. The first phase of drug/xenobiotic metabolism is phase I and is mediated primarily by the cytochrome P450 enzymes. These enzymes catalyze oxidation and reduction reactions. The second phase, phase II, is mediated by enzymes such as glutathione S-transferases, UDP-glucuronosyl transferases, and sulfotransferases; and involves primarily conjugation reactions. During phase III metabolism, the drug/ xenobiotic or drug/xenobiotic conjugates are excreted from the cell via transporter proteins. The enzymes and transporters that are involved in phases I, II, and III metabolism are members of large superfamilies with often overlapping substrate specificities and tissue expression profiles. The highest composite expression levels of these enzymes and transporters are found in the liver with lesser amounts expressed in the gastrointestinal tract, kidney, lung, and brain. The expression levels of drug metabolizing enzymes and transporters are regulated by a number of nuclear receptors, in particular the CAR, PXR, NRF2 and AHR. Each nuclear receptor is capable of upregulating a coordinate set of phase I, II, and III enzymes and transporters that may be distinct or may overlap with that of other nuclear receptors.

Flavonoids are typically absorbed by the small intestine and colon as glycosides.[147] They undergo extensive phase II metabolism in the enterocyte and exit into the bloodstream as sulfates, glucuronides, and methylated metabolites. Within the liver, they are subjected to further phase II metabolism and serve as substrates of the phase III transporters. These transporters play an important role in limiting the bioavailability of flavonoids. While a considerable amount of flavonoid metabolites are excreted in the urine, this level, as well as that subjected to metabolism by the colonic microflora that circulate in the plasma or are sequestered within a particular tissue, is dependent on the flavonoid subclass and the complexity of its structure. The bioavailability of flavonoids is thought to range from 2. 5 to 18. 5% of the consumed flavonoid. Of all the flavonoid subclasses, anthocyanins are thought to have the lowest bioavailability. Thus, the bioavailability of flavonoids is quite low and the levels capable of reaching the plasma and target tissue are subject to host and microbiota expressed enzymes and transporters.

Impact of flavonoids on drug metabolizing enzymes and transporters. Many flavonoids are capable of inhibiting the actions of drug metabolizing enzymes and transporters via competitive inhibition.[148] Thus, high dietary consumption of these flavonoids may enhance the efficacy and/or toxicity of an administered drug. Alternatively, the co-consumption of a cytochrome P450-inhibiting flavonoid with a dietary carcinogen (such as benzo[a]pyrene) would reduce the bioactivation of the carcinogen, decrease formation of carcinogen-DNA adducts, and inhibit formation of the initiated tumor cell.[148] Amongst the best defined interactions between flavonoid-rich foods and drugs is that mediated by grapefruit juice.[149] The consumption of grapefruit juice is known to significantly increase the bioavailability of a wide therapeutic array of drugs. The underlying mechanisms involve irreversible inhibition of phase I enzymes, in particular cytochrome P450 3A and short-lived inhibition of transport proteins. It is thought that the inhibition of cytochrome P450 3A does not involve naringin, the most abundant flavonoid found in grapefruit juice, but primarily involves the furanocoumarins such as bergamottin. However, naringin and its aglycone naringenin are potent inhibitors of transport proteins such as OATP.

Flavonoids in addition to naringin have been examined for their ability to inhibit drug metabolism and thereby alter the bioavailability of drugs and xenobiotics. With respect to inhibition of phase I metabolism, a recent study found that a structurally diverse group of flavonoids were capable of inhibiting cytochrome P450 1A1, 1A2, 1B1, 2C9, and 3A4.[150] However, the extent to which these flavonoids inhibited cytochrome P450 enzyme activity differed, as the underlying mechanisms were dependent on the specific flavonoid and cytochrome P450 involved. Galangin was identified as one of the most potent inhibitors of cytochrome P450 1A1, 1B1, and 3A4. The other flavonoids exhibited varying inhibitory properties on the cytochrome P450s depending on the extent of their hydroxylation at the A-, B-, and C-rings of the flavonoid. Further, the interactions between the flavonoids and the cytochrome P450s occurred at not only the active enzyme site, but within other sites as well. Several flavonoids have also been shown to inhibit phase III metabolism. For example, oral administration of quercetin enhances the bioavailability and corresponding efficacy of the chemotherapeutic drug, paclitaxel, presumably via its competitive inhibition of both cytochrome P450 3A4 and the ABC transporter, P-glycoprotein.[151] However, quercetin can enhance drug metabolism.[152] In a study using human subjects, daily administration of quercetin (500 mg) was found to significantly increase metabolism and reduce the plasma levels of the cytochrome P450 3A probe drug midazolam. Further, the extent to which quercetin induced cytochrome P450 3A-mediated metabolism was dependent on whether or not the individuals harbored CYP3A5*/1* or CYP3A5*1/3* genotypes.

Other flavonoids implicated in the inhibition of P-glycoprotein and phase III metabolism includes genistein, chrysin, hesperetin, naringenin, and epigallocatechin.[153] Similarly, a survey of 56 flavonoids revealed that apigenin, kaempferol, myricetin, and tricetin are relatively potent inhibitors of the ABCG2 transporter.[154] While the extent to which flavonoid-inhibition of drug metabolism and drug transport may be exploited to enhance the efficacy of chemotherapeutic drugs is currently under intense investigation, its role in mediating the chemopreventive actions of flavonoids is not yet clear.

Impact of flavonoids on the regulators of drug metabolizing enzymes and transporters. As mentioned previously, a number of ligand-activated transcription factors regulate the expression levels of drug metabolizing enzymes and transporters.[146] Ligands to these transcription factors include dietary substances, carcinogens, endogenous lipophilic substances and pharmaceutical drugs. By serving as receptor agonists, flavonoids can enhance the elimination of dietary carcinogens, hormones, or drugs, thereby limiting their physiological effects. The extent to which the chemopreventive effects of flavonoids involve enhanced clearance of carcinogens is discussed below in relation to their impact on the AHR pathway. While the regulators of drug metabolizing enzymes and transporters can also modulate many of the tumorigenic pathways discussed above, the discussion below will be limited primarily to their impact of drug and xenobiotic metabolism.

PXR, expressed primarily in the liver and intestine, is activated by a structurally diverse group of substances and regulates its target genes via dimerization with RXR.[155] With respect to phase I metabolizing enzymes, PXR regulates members of the cytochrome P450 3A, 2B, 2C families. With respect to phase II metabolizing enzymes, PXR upregulates glutathione transferase A1 and uridine diphosphate glycosyltransferase 1A3 and 1A6. PXR also regulates several transporters that play important roles in both the cellular influx and efflux of xenobiotics and endogenous substances. Of particular interest is its regulation of MDR (multiple drug resistance 1, also known as P-glycoprotein 1) which plays a key role in mediating a tumor cell's resistance to chemotherapeutics. In addition, PXR is involved in a mutual repressive interaction with NFκB, particularly in the intestine. With respect to flavonoids that activate PXR, chrysin has been shown to activate the PXR pathway and when administered *in vivo*, upregulate the expression of PXR target genes (i.e., cytochrome P450 3A11 and MDR1) in the colon.[156] A recent screen of 40 phytochemicals revealed that tangeretin was amongst the most potent with respect to activation of PXR-signaling.[157]

Like PXR, CAR is a member of the steroid receptor superfamily and regulates its cognate genes via formation of a DNA-binding heterodimer with RXR.[146] Drug metabolizing enzymes and transporters that are upregulated by CAR include cytochrome P450 2B6 (phase I), uridine diphosphate glycosyltransferase B1 and sulfotransferase 1E1 (phase II), and organic anionic transport protein 1B3 (phase III). PXR and CAR regulate overlapping sets of genes involved in phase I, phase II, and phase III metabolism. With respect to flavonoids capable of activating CAR, a survey of flavonoids that included chrysin, baicalein, and galangin revealed that only chrysin was capable of exerting sufficient potency with respect to activating CAR and inducing a CAR (cytochrome P450 2B10) target gene in the liver.[158] Other flavonoids that may act as CAR agonists include ellagic acid[159] and possibly quercetin (and/or its metabolites).[160]

The NRF2 regulated pathway has been described as a "master regulator of a battery of defensive and detoxification genes," given its sensitivity to the presence of reactive oxygen species and electrophiles and the wide variety of protective genes that it regulates.[161–163] Under homeostatic conditions, NRF2 is transcriptionally inactive and associates with its cytosolic partner, KEAP1. Activation of NRF2 by a diverse number of chemical compounds is thought to involve their reaction with sulfhydryl (i.e., cysteine thiols) residues of KEAP1, dissociation of the NRF2-KEAP1 complex, and nuclear translocation of NRF2. Additional modes of NRF2 activation may require MAPK and PIK3CA kinases. With respect to its antioxidant properties, NRF2 regulates genes involved in the synthesis and conjugation of glutathione and antioxidants as discussed above. With respect to drug metabolizing enzymes and transporters, NRF2 upregulates genes such as NAD(P)H dehydrogenase quinone 1 (phase I), a number of glutathione S-transferases (alpha 4, mu 2, and mu 3)(phase 2) and several phase III transporters (e.g., ATP-binding cassette subfamily B and solute carrier 48, member 1). In addition, NRF2 upregulates the expression levels of other transcription factors involved in drug metabolism such as RXRα and the AHR. Examples of flavonoids that have been shown to activate the NRF2 pathway are isorhamnetin,[164] genistein,[165] and luteolin.[166]

The AHR is a member of the basic helix-loop-helix Per-Arnt-Sim family.[146,167] In the presence of its agonists, it translocates into the nucleus, forms a heterodimer with its DNA binding partner ARNT (aryl hydrocarbon receptor nuclear translocator), and binds specific DNA sequences to upregulate its target genes. Genes typically upregulated by the AHR encode cytochrome P450 1A1 and 1B1 enzymes (phase I), glutathione S transferases A1 and A2 and uridine diphosphate glycosyltransferase 1A1 (phase II), and multidrug resistance-associated protein 3 (phase III). The AHR is highly expressed in the liver, lungs, spleen, and kidney. In addition, high expression levels of the AHR and its target gene CYP1B1 are often observed in tumor cells relative to that in the adjacent normal cells. The AHR is probably best recognized for its role in the metabolic activation and excretion of chemical carcinogens such as benzo[a]pyrene and similar environmental polyaromatic hydrocarbons which act as AHR agonists. Some of the activities mediated by the AHR are coordinated with that of NRF2, NFκB, and estrogen receptor α. In addition, the AHR also appears to facilitate additional biological activities of flavonoids including their inhibition of inflammation (discussed in Chapter 3).

Flavonoids appear to be capable of acting as either AHR agonists or antagonists.[167–169] Surveys of a variety of flavonoids have shown that apigenin, galangin, kaempferol, and luteolin can be effective AHR ligands but their abilities to either activate or inhibit the AHR pathway may be cell-specific. Quercetin can activate the AHR, but its actions are indirect and thus should not be considered a true AHR agonist.[170] As AHR agonists, flavonoids exert a protective role by increasing the expression of a number of antioxidant and phase II enzymes. For example, in a mouse model of intestinal carcinogenesis, the administration of dietary AHR agonists inhibited tumor formation.[171] However, in a study of oral tumor cells, either apigenin or kaempferol, both of which act as AHR antagonists,[168,172] inhibited tumor cell growth *in vitro*, but when administered *in vivo*, increased the growth of the explanted tumors.[173] In human subjects, the ability of flavonoids to activate the AHR has been examined in a flavonoid-supplementation study.[174] Here, a one-month increase in the consumption of flavonoids appeared to correspond to an increase in CYP1A1 expression which is indicative of AHR activation. In addition, increased expression levels of DNA repair genes were observed, but not at a level of significance. Thus, the overall impact of AHR activation with respect to the chemopreventive properties of flavonoids is yet to be determined.

References

1. Hanahan D, Weinberg RA. Hallmarks of cancer: The next generation. *Cell.* Mar 4, 2011;144(5):646–674.
2. Hynes NE, MacDonald G. ErbB receptors and signaling pathways in cancer. *Curr Opin Cell Biol.* Apr 2009;21(2):177–184.
3. Suh Y, Afaq F, Johnson JJ, Mukhtar H. A plant flavonoid fisetin induces apoptosis in colon cancer cells by inhibition of COX2 and Wnt/EGFR/NF-kappaB-signaling pathways. *Carcinogenesis.* Feb 2009;30(2):300–307.

4. Yang J, Li Q, Zhou XD, Kolosov VP, Perelman JM. Naringenin attenuates mucous hypersecretion by modulating reactive oxygen species production and inhibiting NF-kappaB activity via EGFR-PI3K-Akt/ERK MAPKinase signaling in human airway epithelial cells. *Mol Cell Biochem.* May 2011; 351(1–2):29–40.

5. Masuelli L, Marzocchella L, Quaranta A, *et al.* Apigenin induces apoptosis and impairs head and neck carcinomas EGFR/ErbB2 signaling. *Front Biosci (Landmark Ed).* 2011;16:1060–1068.

6. Jung JH, Lee JO, Kim JH, *et al.* Quercetin suppresses HeLa cell viability via AMPK-induced HSP70 and EGFR down-regulation. *J Cell Physiol.* May 2010;223(2):408–414.

7. Potapovich AI, Lulli D, Fidanza P, *et al.* Plant polyphenols differentially modulate inflammatory responses of human keratinocytes by interfering with activation of transcription factors NFkappaB and AhR and EGFR-ERK pathway. *Toxicol Appl Pharmacol.* Sep 1, 2011;255(2):138–149.

8. Choi S, Lim TG, Hwang MK, *et al.* Rutin inhibits B[a]PDE-induced cyclo-oxygenase-2 expression by targeting EGFR kinase activity. *Biochem Pharmacol.* Nov 15, 2013;86(10):1468–1475.

9. Fang J, Zhou Q, Shi XL, Jiang BH. Luteolin inhibits insulin-like growth factor 1 receptor signaling in prostate cancer cells. *Carcinogenesis.* Mar 2007;28(3):713–723.

10. Kang Y, Park MA, Heo SW, *et al.* The radio-sensitizing effect of xanthohumol is mediated by STAT3 and EGFR suppression in doxorubicin-resistant MCF-7 human breast cancer cells. *Biochim Biophys Acta.* Mar 2013; 1830(3):2638–2648.

11. Jeong JH, An JY, Kwon YT, Li LY, Lee YJ. Quercetin-induced ubiquitination and down-regulation of Her-2/neu. *J Cell Biochem.* Oct 1, 2008; 105(2):585–595.

12. Shukla S, MacLennan GT, Fu P, Gupta S. Apigenin attenuates insulin-like growth factor-I signaling in an autochthonous mouse prostate cancer model. *Pharm Res.* Jun 2012;29(6):1506–1517.

13. Koul HK, Pal M, Koul S. Role of p38 MAP Kinase Signal Transduction in Solid Tumors. *Genes Cancer.* Sep 2013;4(9–10):342–359.

14. Deschenes-Simard X, Kottakis F, Meloche S, Ferbeyre G. ERKs in cancer: Friends or foes? *Cancer Res.* Jan 15, 2014;74(2):412–419.

15. McCubrey JA, Steelman LS, Chappell WH, *et al.* Roles of the Raf/MEK/ERK pathway in cell growth, malignant transformation and drug resistance. *Biochim Biophys Acta.* Aug 2007;1773(8):1263–1284.

16. Martini M, De Santis MC, Braccini L, Gulluni F, Hirsch E. PI3K/AKT signaling pathway and cancer: An updated review. *Ann Med.* Jun 5:1–12.

17. Park SH, Kim JH, Lee DH, *et al.* Luteolin 8-C-beta-fucopyranoside inhibits invasion and suppresses TPA-induced MMP-9 and IL-8 via ERK/AP-1 and ERK/NF-kappaB signaling in MCF-7 breast cancer cells. *Biochimie.* Nov 2013;95(11):2082–2090.

18. Chen HJ, Lin CM, Lee CY, *et al.* Kaempferol suppresses cell metastasis via inhibition of the ERK-p38-JNK and AP-1 signaling pathways in U-2 OS human osteosarcoma cells. *Oncol Rep.* Aug 2013;30(2):925–932.

19. Yuan L, Wang J, Xiao H, Wu W, Wang Y, Liu X. MAPK signaling pathways regulate mitochondrial-mediated apoptosis induced by isoorientin in human hepatoblastoma cancer cells. *Food Chem Toxicol.* Mar 2013;53:62–68.

20. Liang RR, Zhang S, Qi JA, *et al.* Preferential inhibition of hepatocellular carcinoma by the flavonoid Baicalein through blocking MEK-ERK signaling. *Int J Oncol.* Sep 2012;41(3):969–978.

21. Lu Z, Lu N, Li C, *et al.* Oroxylin A inhibits matrix metalloproteinase-2/9 expression and activation by up-regulating tissue inhibitor of metalloproteinase-2 and suppressing the ERK1/2 signaling pathway. *Toxicol Lett.* Mar 25;209(3):211–220.

22. Ding M, Zhao J, Bowman L, Lu Y, Shi X. Inhibition of AP-1 and MAPK signaling and activation of Nrf2/ARE pathway by quercitrin. *Int J Oncol.* Jan 2010;36(1):59–67.

23. Lee KW, Kang NJ, Heo YS, *et al.* Raf and MEK protein kinases are direct molecular targets for the chemopreventive effect of quercetin, a major flavonol in red wine. *Cancer Res.* Feb 1, 2008;68(3):946–955.

24. Lee KW, Kang NJ, Rogozin EA, *et al.* Myricetin is a novel natural inhibitor of neoplastic cell transformation and MEK1. *Carcinogenesis.* Sep 2007;28(9):1918–1927.

25. Zhai X, Lin M, Zhang F, *et al.* Dietary flavonoid genistein induces Nrf2 and phase II detoxification gene expression via ERKs and PKC pathways and protects against oxidative stress in Caco-2 cells. *Mol Nutr Food Res.* Feb;57(2):249–259.

26. Androutsopoulos VP, Spandidos DA. The flavonoids diosmetin and luteolin exert synergistic cytostatic effects in human hepatoma HepG2 cells via

CYP1A-catalyzed metabolism, activation of JNK and ERK and P53/P21 up-regulation. *J Nutr Biochem.* Feb;24(2):496–504.

27. Rodriguez-Ramiro I, Ramos S, Bravo L, Goya L, Martin MA. Procyanidin B2 induces Nrf2 translocation and glutathione S-transferase P1 expression via ERKs and p38-MAPK pathways and protect human colonic cells against oxidative stress. *Eur J Nutr.* Oct 2012;51(7):881–892.

28. Llorens F, Miro FA, Casanas A, *et al.* Unbalanced activation of ERK1/2 and MEK1/2 in apigenin-induced HeLa cell death. *Exp Cell Res.* Sep 10, 2004; 299(1):15–26.

29. Chou RH, Hsieh SC, Yu YL, Huang MH, Huang YC, Hsieh YH. Fisetin inhibits migration and invasion of human cervical cancer cells by down-regulating urokinase plasminogen activator expression through suppressing the p38 MAPK-dependent NF-kappaB signaling pathway. *PLoS One.* 2013;8(8):e71983.

30. Yuan L, Wu Y, Ren X, Liu Q, Wang J, Liu X. Isoorientin attenuates lipopoly-saccharide-induced pro-inflammatory responses through down-regulation of ROS-related MAPK/NF-kappaB signaling pathway in BV-2 microglia. *Mol Cell Biochem.* Jan 2013;386(1–2):153–165.

31. Weng CJ, Chen MJ, Yeh CT, Yen GC. Hepatoprotection of quercetin against oxidative stress by induction of metallothionein expression through activating MAPK and PI3K pathways and enhancing Nrf2 DNA-binding activity. *N Biotechnol.* Oct 2011;28(6):767–777.

32. Torres F, Quintana J, Diaz JG, Carmona AJ, Estevez F. Trifolin acetate-induced cell death in human leukemia cells is dependent on caspase-6 and activates the MAPK pathway. *Apoptosis.* May 2008;13(5):716–728.

33. Galluzzo P, Martini C, Bulzomi P, *et al.* Quercetin-induced apoptotic cascade in cancer cells: Antioxidant versus estrogen receptor alpha-dependent mechanisms. *Mol Nutr Food Res.* Jun 2009;53(6):699–708.

34. Fong Y, Shen KH, Chiang TA, Shih YW. Acacetin inhibits TPA-induced MMP-2 and u-PA expressions of human lung cancer cells through inactivating JNK signaling pathway and reducing binding activities of NF-kappaB and AP-1. *J Food Sci.* Jan–Feb 2010;75(1):H30–38.

35. Baek S, Kang NJ, Popowicz GM, *et al.* Structural and functional analysis of the natural JNK1 inhibitor quercetagetin. *J Mol Biol.* Jan 23;425(2):411–423.

36. Bajer MM, Kunze MM, Blees JS, *et al.* Characterization of pomiferin triacetate as a novel mTOR and translation inhibitor. *Biochem Pharmacol.* Apr 1, 2014; 88(3):313–321.

37. Xia J, Guo S, Fang T, *et al.* Dihydromyricetin induces autophagy in HepG2 cells involved in inhibition of mTOR and regulating its upstream pathways. *Food Chem Toxicol.* Apr 2014;66:7–13.

38. Jin YM, Xu TM, Zhao YH, Wang YC, Cui MH. *In vitro* and *in vivo* anti-cancer activity of formononetin on human cervical cancer cell line HeLa. *Tumour Biol.* Mar 2014;35(3):2279–2284.

39. Pozsgai E, Bellyei S, Cseh A, *et al.* Quercetin increases the efficacy of glioblastoma treatment compared to standard chemoradiotherapy by the suppression of PI-3-kinase-Akt pathway. *Nutr Cancer.* 2013;65(7): 1059–1066.

40. Gao AM, Ke ZP, Shi F, Sun GC, Chen H. Chrysin enhances sensitivity of BEL-7402/ADM cells to doxorubicin by suppressing PI3K/Akt/Nrf2 and ERK/Nrf2 pathway. *Chem Biol Interact.* Oct 25, 2013;206(1):100–108.

41. Liu S, Li H, Chen L, *et al.* (–)-Epigallocatechin-3-gallate inhibition of Epstein-Barr virus spontaneous lytic infection involves ERK1/2 and PI3-K/Akt signaling in EBV-positive cells. *Carcinogenesis.* Mar 2013;34(3):627–637.

42. Cordero-Herrera I, Martin MA, Bravo L, Goya L, Ramos S. Epicatechin gallate induces cell death via p53 activation and stimulation of p38 and JNK in human colon cancer SW480 cells. *Nutr Cancer.* 2013;65(5):718–728.

43. Lin C, Tsai SC, Tseng MT, *et al.* AKT serine/threonine protein kinase modulates baicalin-triggered autophagy in human bladder cancer T24 cells. *Int J Oncol.* Mar 2013;42(3):993–1000.

44. Wang KL, Hsia SM, Chan CJ, *et al.* Inhibitory effects of isoliquiritigenin on the migration and invasion of human breast cancer cells. *Expert Opin Ther Targets.* Apr 2013;17(4):337–349.

45. Yang Y, Li XJ, Chen Z, *et al.* Wogonin induced calreticulin/annexin A1 exposure dictates the immunogenicity of cancer cells in a PERK/AKT dependent manner. *PLoS One.* 2012;7(12):e50811.

46. Kim HY, Jung SK, Byun S, *et al.* Raf and PI3K are the molecular targets for the anti-metastatic effect of luteolin. *Phytother Res.* Oct 2013;27(10):1481–1488.

47. Oi N, Chen H, Ok Kim M, Lubet RA, Bode AM, Dong Z. Taxifolin suppresses UV-induced skin carcinogenesis by targeting EGFR and PI3K. *Cancer Prev Res (Phila).* Sep 2012;5(9):1103–1114.

48. Khan N, Adhami VM, Afaq F, Mukhtar H. Butein induces apoptosis and inhibits prostate tumor growth *in vitro* and *in vivo. Antioxid Redox Signal.* Jun 1, 2012;16(11):1195–1204.

49. Wang Y, Zhao Y, Liu Y, Tian L, Jin D. Chamaejasmine inactivates Akt to trigger apoptosis in human HEp-2 larynx carcinoma cells. *Molecules.* 16(10):8152–8164.

50. Kang NJ, Jung SK, Lee KW, Lee HJ. Myricetin is a potent chemopreventive phytochemical in skin carcinogenesis. *Ann N Y Acad Sci.* Jul 2011;1229:124–132.

51. Khan N, Afaq F, Khusro FH, Mustafa Adhami V, Suh Y, Mukhtar H. Dual inhibition of phosphatidylinositol 3-kinase/Akt and mammalian target of rapamycin signaling in human nonsmall cell lung cancer cells by a dietary flavonoid fisetin. *Int J Cancer.* Apr 1, 2012;130(7):1695–1705.

52. Lee YC, Cheng TH, Lee JS, *et al.* Nobiletin, a citrus flavonoid, suppresses invasion and migration involving FAK/PI3K/Akt and small GTPase signals in human gastric adenocarcinoma AGS cells. *Mol Cell Biochem.* Jan 2011;347(1–2):103–115.

53. Arafa el SA, Zhu Q, Barakat BM, *et al.* Tangeretin sensitizes cisplatin-resistant human ovarian cancer cells through downregulation of phosphoinositide 3-kinase/Akt signaling pathway. *Cancer Res.* Dec 1, 2009;69(23):8910–8917.

54. Kwon JY, Lee KW, Kim JE, *et al.* Delphinidin suppresses ultraviolet B-induced cyclooxygenases-2 expression through inhibition of MAPKK4 and PI-3 kinase. *Carcinogenesis.* Nov 2009;30(11):1932–1940.

55. Anastasius N, Boston S, Lacey M, Storing N, Whitehead SA. Evidence that low-dose, long-term genistein treatment inhibits oestradiol-stimulated growth in MCF-7 cells by down-regulation of the PI3-kinase/Akt signalling pathway. *J Steroid Biochem Mol Biol.* Aug 2009;116(1–2):50–55.

56. Zhang Q, Liu J, Liu B, *et al.* Dihydromyricetin promotes hepatocellular carcinoma regression via a p53 activation-dependent mechanism. *Sci Rep.* 2014;4:4628.

57. Li J, Qu W, Cheng Y, *et al.* The Inhibitory Effect of Intravesical Fisetin against Bladder Cancer by Induction of p53 and Down-Regulation of NF-kappa B Pathways in a Rat Bladder Carcinogenesis Model. *Basic Clin Pharmacol Toxicol.* Mar 19, 2014.

58. Lee Y, Sung B, Kang YJ, *et al.* Apigenin-induced apoptosis is enhanced by inhibition of autophagy formation in HCT116 human colon cancer cells. *Int J Oncol.* May 2014;44(5):1599–1606.

59. Kim H, Moon JY, Ahn KS, Cho SK. Quercetin induces mitochondrial mediated apoptosis and protective autophagy in human glioblastoma U373MG cells. *Oxid Med Cell Longev.* 2013;2013:596496.

60. Kim MS, Bak Y, Park YS, *et al*. Wogonin induces apoptosis by suppressing E6 and E7 expressions and activating intrinsic signaling pathways in HPV-16 cervical cancer cells. *Cell Biol Toxicol*. Aug 2013;29(4):259–272.

61. Arul D, Subramanian P. Naringenin (citrus flavonone) induces growth inhibition, cell cycle arrest and apoptosis in human hepatocellular carcinoma cells. *Pathology oncology research: POR*. Oct 2013;19(4):763–770.

62. Wang TT, Wang SK, Huang GL, Sun GJ. Luteolin induced-growth inhibition and apoptosis of human esophageal squamous carcinoma cell line Eca109 cells *in vitro*. *Asian Pacific journal of cancer prevention: APJCP*. 2012;13(11):5455–5461.

63. Xu F, Zang J, Chen D, *et al*. Neohesperidin induces cellular apoptosis in human breast adenocarcinoma MDA-MB-231 cells via activating the Bcl-2/Bax-mediated signaling pathway. *Natural product communications*. Nov 2012;7(11):1475–1478.

64. Kim JH, Lee JO, Lee SK, *et al*. Celastrol suppresses breast cancer MCF-7 cell viability via the AMP-activated protein kinase (AMPK)-induced p53-polo like kinase 2 (PLK-2) pathway. *Cellular signalling*. Apr 2013; 25(4):805–813.

65. Cheng YH, Li LA, Lin P, *et al*. Baicalein induces G1 arrest in oral cancer cells by enhancing the degradation of cyclin D1 and activating AhR to decrease Rb phosphorylation. *Toxicol Appl Pharmacol*. Sep 15, 2012; 263(3):360–367.

66. Maso V, Calgarotto AK, Franchi Jr. GC, *et al*. Multitarget effects of quercetin in leukemia. *Cancer Prev Res (Phila)*. Dec 2014;7(12):1240–1250.

67. Ong CS, Zhou J, Ong CN, Shen HM. Luteolin induces G1 arrest in human nasopharyngeal carcinoma cells via the Akt-GSK-3beta-Cyclin D1 pathway. *Cancer letters*. Dec 8, 2010;298(2):167–175.

68. Kim DC, Ramachandran S, Baek SH, *et al*. Induction of growth inhibition and apoptosis in human uterine leiomyoma cells by isoliquiritigenin. *Reproductive sciences*. Jul 2008;15(6):552–558.

69. Shukla S, Gupta S. Apigenin-induced cell cycle arrest is mediated by modulation of MAPK, PI3K-Akt, and loss of cyclin D1 associated retinoblastoma dephosphorylation in human prostate cancer cells. *Cell cycle*. May 2, 2007; 6(9):1102–1114.

70. Kim HR, Park CG, Jung JY. Acacetin (5,7-dihydroxy-4′-methoxyflavone) exhibits *in vitro* and *in vivo* anticancer activity through the suppression of

NF-kappaB/Akt signaling in prostate cancer cells. *International journal of molecular medicine.* Feb 2014;33(2):317–324.

71. Wang Z, Lu W, Li Y, Tang B. Alpinetin promotes Bax translocation, induces apoptosis through the mitochondrial pathway and arrests human gastric cancer cells at the G2/M phase. *Mol Med Rep.* Mar 2013;7(3):915–920.

72. Lin CC, Chuang YJ, Yu CC, *et al.* Apigenin induces apoptosis through mito-chondrial dysfunction in U-2 OS human osteosarcoma cells and inhibits osteosarcoma xenograft tumor growth *in vivo. J Agric Food Chem.* Nov 14, 2012;60(45):11395–11402.

73. Zhang Y, Song L, Cai L, Wei R, Hu H, Jin W. Effects of baicalein on apop-tosis, cell cycle arrest, migration and invasion of osteosarcoma cells. *Food Chem Toxicol.* Mar 2013;53:325–333.

74. Li Y, Ma C, Qian M, Wen Z, Jing H, Qian D. Butein induces cell apoptosis and inhibition of cyclooxygenase2 expression in A549 lung cancer cells. *Mol Med Rep.* Feb 2014;9(2):763–767.

75. Pal HC, Sharma S, Elmets CA, Athar M, Afaq F. Fisetin inhibits growth, induces G(2) /M arrest and apoptosis of human epidermoid carcinoma A431 cells: Role of mitochondrial membrane potential disruption and consequent caspases activation. *Experimental dermatology.* Jul 2013;22(7):470–475.

76. Zhang W, Lan Y, Huang Q, Hua Z. Galangin induces B16F10 melanoma cell apoptosis via mitochondrial pathway and sustained activation of p38 MAPK. *Cytotechnology.* May 2013;65(3):447–455.

77. Qiao Y, Xiang Q, Yuan L, Xu L, Liu Z, Liu X. Herbacetin induces apoptosis in HepG2 cells: Involvements of ROS and PI3K/Akt pathway. *Food Chem Toxicol.* Jan 2013;51:426–433.

78. Alshatwi AA, Ramesh E, Periasamy VS, Subash-Babu P. The apoptotic effect of hesperetin on human cervical cancer cells is mediated through cell cycle arrest, death receptor, and mitochondrial pathways. *Fundamental & clinical pharmacology.* Dec 2013;27(6):581–592.

79. Gao H, Wang H, Peng J. Hispidulin induces apoptosis through mitochon-drial dysfunction and inhibition of P13k/Akt signalling pathway in HepG2 cancer cells. *Cell biochemistry and biophysics.* May 2014;69(1):27–34.

80. Li Y, Zhao H, Wang Y, *et al.* Isoliquiritigenin induces growth inhibition and apoptosis through downregulating arachidonic acid metabolic network and the deactivation of PI3K/Akt in human breast cancer. *Toxicol Appl Pharmacol.* Oct 1, 2013;272(1):37–48.

81. Verschooten L, Barrette K, Van Kelst S, *et al*. Autophagy inhibitor chloro-quine enhanced the cell death inducing effect of the flavonoid luteolin in metastatic squamous cell carcinoma cells. *PLoS One*. 2012;7(10):e48264.

82. Luo H, Rankin GO, Li Z, Depriest L, Chen YC. Kaempferol induces apoptosis in ovarian cancer cells through activating p53 in the intrinsic pathway. *Food chemistry*. Sep 15, 2011;128(2):513–519.

83. Shu G, Yang J, Zhao W, *et al*. Kurarinol induces hepatocellular carcinoma cell apoptosis through suppressing cellular signal transducer and activator of transcription 3 signaling. *Toxicol Appl Pharmacol*. Dec 1, 2014;281(2):157–165.

84. Xu R, Zhang Y, Ye X, *et al*. Inhibition effects and induction of apoptosis of flavonoids on the prostate cancer cell line PC-3 *in vitro*. *Food chemistry*. May 1, 2013;138(1):48–53.

85. Chen H, Miao Q, Geng M, *et al*. Anti-tumor effect of rutin on human neuroblastoma cell lines through inducing G2/M cell cycle arrest and promoting apoptosis. *The Scientific World Journal*. 2013;2013:269165.

86. Nicolini F, Burmistrova O, Marrero MT, *et al*. Induction of G2/M phase arrest and apoptosis by the flavonoid tamarixetin on human leukemia cells. *Mol Carcinog*. Dec 2014;53(12):939–950.

87. Zhang L, Wang H, Cong Z, *et al*. Wogonoside induces autophagy-related apoptosis in human glioblastoma cells. *Oncol Rep*. Sep 2014;32(3):1179–1187.

88. Wang Z, Jiang C, Chen W, *et al*. Baicalein induces apoptosis and autophagy via endoplasmic reticulum stress in hepatocellular carcinoma cells. *BioMed research international*. 2014;2014:732516.

89. Suh Y, Afaq F, Khan N, Johnson JJ, Khusro FH, Mukhtar H. Fisetin induces autophagic cell death through suppression of mTOR signaling pathway in prostate cancer cells. *Carcinogenesis*. Aug 2010;31(8):1424–1433.

90. Wen M, Wu J, Luo H, Zhang H. Galangin induces autophagy through upregulation of p53 in HepG2 cells. *Pharmacology*. 2012;89(5–6):247–255.

91. Ni F, Gong Y, Li L, Abdolmaleky HM, Zhou JR. Flavonoid ampelopsin inhibits the growth and metastasis of prostate cancer *in vitro* and in mice. *PLoS One*. 2012;7(6):e38802.

92. Mafuvadze B, Liang Y, Besch-Williford C, Zhang X, Hyder SM. Apigenin induces apoptosis and blocks growth of medroxyprogesterone acetate-dependent BT-474 xenograft tumors. *Hormones & cancer*. Aug 2012; 3(4):160–171.

93. Chen D, Landis-Piwowar KR, Chen MS, Dou QP. Inhibition of proteasome activity by the dietary flavonoid apigenin is associated with growth inhibition in cultured breast cancer cells and xenografts. *Breast cancer research: BCR.* 2007;9(6):R80.

94. Pandey M, Kaur P, Shukla S, Abbas A, Fu P, Gupta S. Plant flavone apigenin inhibits HDAC and remodels chromatin to induce growth arrest and apoptosis in human prostate cancer cells: *In vitro* and *in vivo* study. *Mol Carcinog.* Dec;51(12):952–962.

95. Dang Q, Song W, Xu D, *et al.* Kaempferol suppresses bladder cancer tumor growth by inhibiting cell proliferation and inducing apoptosis. *Mol Carcinog.* Apr 2, 2014.

96. Phillips PA, Sangwan V, Borja-Cacho D, Dudeja V, Vickers SM, Saluja AK. Myricetin induces pancreatic cancer cell death via the induction of apoptosis and inhibition of the phosphatidylinositol 3-kinase (PI3K) signaling pathway. *Cancer letters.* Sep 28, 2011;308(2):181–188.

97. Hu R, Chen N, Yao J, *et al.* The role of Nrf2 and apoptotic signaling pathways in oroxylin A-mediated responses in HCT-116 colorectal adenocarcinoma cells and xenograft tumors. *Anti-cancer drugs.* Jul 2012;23(6):651–658.

98. Angst E, Park JL, Moro A, *et al.* The flavonoid quercetin inhibits pancreatic cancer growth *in vitro* and *in vivo. Pancreas.* Mar 2013;42(2):223–229.

99. Casella ML, Parody JP, Ceballos MP, *et al.* Quercetin prevents liver carcinogenesis by inducing cell cycle arrest, decreasing cell proliferation and enhancing apoptosis. *Mol Nutr Food Res.* Feb 2014;58(2):289–300.

100. Kim DH, Hossain MA, Kang YJ, *et al.* Baicalein, an active component of Scutellaria baicalensis Georgi, induces apoptosis in human colon cancer cells and prevents AOM/DSS-induced colon cancer in mice. *Int J Oncol.* Nov 2013;43(5):1652–1658.

101. Jayasooriya RG, Kang SH, Kang CH, *et al.* Apigenin decreases cell viability and telomerase activity in human leukemia cell lines. *Food Chem Toxicol.* Aug 2012;50(8):2605–2611.

102. Moon DO, Kim MO, Lee JD, Choi YH, Kim GY. Butein suppresses c-Myc-dependent transcription and Akt-dependent phosphorylation of hTERT in human leukemia cells. *Cancer Letters.* Dec 28, 2009;286(2):172–179.

103. Wang X, Hao MW, Dong K, Lin F, Ren JH, Zhang HZ. Apoptosis induction effects of EGCG in laryngeal squamous cell carcinoma cells through telomerase repression. *Archives of pharmacal research.* Sep 2009;32(9):1263–1269.

104. Khaw AK, Yong JW, Kalthur G, Hande MP. Genistein induces growth arrest and suppresses telomerase activity in brain tumor cells. *Genes, chromosomes & cancer.* Oct 2012;51(10):961–974.

105. Avci CB, Yilmaz S, Dogan ZO, *et al.* Quercetin-induced apoptosis involves increased hTERT enzyme activity of leukemic cells. *Hematology.* Sep 2011;16(5):303–307.

106. Huang ST, Wang CY, Yang RC, Chu CJ, Wu HT, Pang JH. Wogonin, an active compound in Scutellaria baicalensis, induces apoptosis and reduces telomerase activity in the HL-60 leukemia cells. *Phytomedicine: International journal of phytotherapy and phytopharmacology.* Jan 2010;17(1):47–54.

107. Naasani I, Oh-Hashi F, Oh-Hara T, *et al.* Blocking telomerase by dietary polyphenols is a major mechanism for limiting the growth of human cancer cells *in vitro* and *in vivo. Cancer Res.* Feb 15, 2003;63(4):824–830.

108. Loi M, Di Paolo D, Becherini P, *et al.* The use of the orthotopic model to validate antivascular therapies for cancer. *The International journal of developmental biology.* 2011;55(4–5):547–555.

109. Bruno A, Pagani A, Pulze L, *et al.* Orchestration of angiogenesis by immune cells. *Frontiers in oncology.* 2014;4:131.

110. Nicosia RF. The aortic ring model of angiogenesis: A quarter century of search and discovery. *Journal of cellular and molecular medicine.* Oct 2009;13(10):4113–4136.

111. Littleton RM, Hove JR. Zebrafish: A nontraditional model of traditional medicine. *Journal of ethnopharmacology.* Feb 13, 2013;145(3):677–685.

112. Kain KH, Miller JW, Jones-Paris CR, *et al.* The chick embryo as an expanding experimental model for cancer and cardiovascular research. *Developmental dynamics: An official publication of the American Association of Anatomists.* Feb 2014;243(2):216–228.

113. Ling Y, Chen Y, Chen P, *et al.* Baicalein potently suppresses angiogenesis induced by vascular endothelial growth factor through the p53/Rb signaling pathway leading to G1/S cell cycle arrest. *Experimental biology and medicine.* Jul 2011;236(7):851–858.

114. Hsu YL, Wu LY, Hou MF, *et al.* Glabridin, an isoflavan from licorice root, inhibits migration, invasion and angiogenesis of MDA-MB-231 human breast adenocarcinoma cells by inhibiting focal adhesion kinase/Rho signaling pathway. *Mol Nutr Food Res.* Feb 2011;55(2):318–327.

115. He L, Wu Y, Lin L, *et al.* Hispidulin, a small flavonoid molecule, suppresses the angiogenesis and growth of human pancreatic cancer by targeting vascular endothelial growth factor receptor 2-mediated PI3K/Akt/mTOR signaling pathway. *Cancer science.* Jan 2011;102(1):219–225.

116. Varinska L, van Wijhe M, Belleri M, *et al.* Anti-angiogenic activity of the flavonoid precursor 4-hydroxychalcone. *Eur J Pharmacol.* Sep 15, 2012; 691(1–3):125–133.

117. Wang Z, Wang N, Han S, *et al.* Dietary compound isoliquiritigenin inhibits breast cancer neoangiogenesis via VEGF/VEGFR-2 signaling pathway. *PLoS One.* 2013;8(7):e68566.

118. Lam KH, Alex D, Lam IK, Tsui SK, Yang ZF, Lee SM. Nobiletin, a polymethoxylated flavonoid from citrus, shows anti-angiogenic activity in a zebrafish *in vivo* model and HUVEC *in vitro* model. *J Cell Biochem.* Nov 2011;112(11):3313–3321.

119. Zhao D, Qin C, Fan X, Li Y, Gu B. Inhibitory effects of quercetin on angiogenesis in larval zebrafish and human umbilical vein endothelial cells. *Eur J Pharmacol.* Jan 15, 2014;723:360–367.

120. Lin CM, Chang H, Chen YH, Wu IH, Chiu JH. Wogonin inhibits IL-6-induced angiogenesis via down-regulation of VEGF and VEGFR-1, not VEGFR-2. *Planta medica.* Nov 2006;72(14):1305–1310.

121. Pratheeshkumar P, Son YO, Budhraja A, *et al.* Luteolin inhibits human prostate tumor growth by suppressing vascular endothelial growth factor receptor 2-mediated angiogenesis. *PLoS One.* 2012;7(12):e52279.

122. Fang J, Zhou Q, Liu LZ, *et al.* Apigenin inhibits tumor angiogenesis through decreasing HIF-1alpha and VEGF expression. *Carcinogenesis.* Apr 2007; 28(4):858–864.

123. Luo H, Rankin GO, Liu L, Daddysman MK, Jiang BH, Chen YC. Kaempferol inhibits angiogenesis and VEGF expression through both HIF dependent and independent pathways in human ovarian cancer cells. *Nutr Cancer.* 2009;61(4):554–563.

124. Lam IK, Alex D, Wang YH, *et al. In vitro* and *in vivo* structure and activity relationship analysis of polymethoxylated flavonoids: Identifying sinensetin as a novel antiangiogenesis agent. *Mol Nutr Food Res.* Jun 2012;56(6): 945–956.

125. Liu LZ, Fang J, Zhou Q, Hu X, Shi X, Jiang BH. Apigenin inhibits expression of vascular endothelial growth factor and angiogenesis in human lung

cancer cells: Implication of chemoprevention of lung cancer. *Molecular pharmacology.* Sep 2005;68(3):635–643.

126. Lamy S, Akla N, Ouanouki A, Lord-Dufour S, Beliveau R. Diet-derived polyphenols inhibit angiogenesis by modulating the interleukin-6/STAT3 pathway. *Exp Cell Res.* Aug 1, 2012;318(13):1586–1596.

127. Gupta GP, Massague J. Cancer metastasis: Building a framework. *Cell.* Nov 17, 2006;127(4):679–695.

128. McAllister SS, Weinberg RA. The tumour-induced systemic environment as a critical regulator of cancer progression and metastasis. *Nature cell biology.* Aug 2014;16(8):717–727.

129. Chunhua L, Donglan L, Xiuqiong F, *et al.* Apigenin up-regulates transgelin and inhibits invasion and migration of colorectal cancer through decreased phosphorylation of AKT. *J Nutr Biochem.* Oct 2013;24(10):1766–1775.

130. He J, Xu Q, Wang M, *et al.* Oral Administration of Apigenin Inhibits Metastasis through AKT/P70S6K1/MMP-9 Pathway in Orthotopic Ovarian Tumor Model. *International journal of molecular sciences.* 2012;13(6): 7271–7282.

131. Lirdprapamongkol K, Sakurai H, Abdelhamed S, *et al.* A flavonoid chrysin suppresses hypoxic survival and metastatic growth of mouse breast cancer cells. *Oncol Rep.* Nov 2013;30(5):2357–2364.

132. Boreddy SR, Srivastava SK. Deguelin suppresses pancreatic tumor growth and metastasis by inhibiting epithelial-to-mesenchymal transition in an orthotopic model. *Oncogene.* Aug 22, 2013;32(34):3980–3991.

133. Zhang W, Tang B, Huang Q, Hua Z. Galangin inhibits tumor growth and metastasis of B16F10 melanoma. *J Cell Biochem.* Jan 2013;114(1):152–161.

134. Zheng H, Li Y, Wang Y, *et al.* Downregulation of COX-2 and CYP 4A signaling by isoliquiritigenin inhibits human breast cancer metastasis through preventing anoikis resistance, migration and invasion. *Toxicol Appl Pharmacol.* Oct 1, 2014;280(1):10–20.

135. Ruan JS, Liu YP, Zhang L, *et al.* Luteolin reduces the invasive potential of malignant melanoma cells by targeting beta3 integrin and the epithelial-mesenchymal transition. *Acta pharmacologica Sinica.* Oct 2012; 33(10): 1325–1331.

136. Fan SH, Wang YY, Lu J, *et al.* Luteoloside suppresses proliferation and metastasis of hepatocellular carcinoma cells by inhibition of NLRP3 inflammasome. *PLoS One.* 2014;9(2):e89961.

137. Lou C, Zhang F, Yang M, *et al.* Naringenin decreases invasiveness and metastasis by inhibiting TGF-beta-induced epithelial to mesenchymal transition in pancreatic cancer cells. *PLoS One.* 2012;7(12):e50956.

138. Wu K, Ning Z, Zeng J, *et al.* Silibinin inhibits beta-catenin/ZEB1 signaling and suppresses bladder cancer metastasis via dual-blocking epithelial-mesenchymal transition and stemness. *Cellular signalling.* Dec 2013; 25(12):2625–2633.

139. Lushchak VI. Free radicals, reactive oxygen species, oxidative stress and its classification. *Chem Biol Interact.* Oct 28, 2014;224C:164–175.

140. Holmstrom KM, Finkel T. Cellular mechanisms and physiological consequences of redox-dependent signalling. *Nature reviews. Molecular cell biology.* Jun 2014;15(6):411–421.

141. Rahal A, Kumar A, Singh V, *et al.* Oxidative stress, prooxidants, and antioxidants: The interplay. *BioMed research international.* 2014;2014:761264.

142. Sayin VI, Ibrahim MX, Larsson E, Nilsson JA, Lindahl P, Bergo MO. Antioxidants accelerate lung cancer progression in mice. *Science translational medicine.* Jan 29, 2014;6(221):221ra215.

143. Fraga CG, Galleano M, Verstraeten SV, Oteiza PI. Basic biochemical mechanisms behind the health benefits of polyphenols. *Molecular aspects of medicine.* Dec 2010;31(6):435–445.

144. Mouradov A, Spangenberg G. Flavonoids: A metabolic network mediating plants adaptation to their real estate. *Frontiers in plant science.* 2014;5:620.

145. Cushnie TP, Lamb AJ. Recent advances in understanding the antibacterial properties of flavonoids. *International journal of antimicrobial agents.* Aug 2011;38(2):99–107.

146. Kohle C, Bock KW. Coordinate regulation of human drug-metabolizing enzymes, and conjugate transporters by the Ah receptor, pregnane X receptor and constitutive androstane receptor. *Biochem Pharmacol.* Feb 15, 2009;77(4):689–699.

147. Del Rio D, Rodriguez-Mateos A, Spencer JP, Tognolini M, Borges G, Crozier A. Dietary (poly)phenolics in human health: Structures, bioavailability, and evidence of protective effects against chronic diseases. *Antioxid Redox Signal.* May 10, 2013;18(14):1818–1892.

148. Moon YJ, Wang X, Morris ME. Dietary flavonoids: Effects on xenobiotic and carcinogen metabolism. *Toxicology in vitro: An international journal published in association with BIBRA.* Mar 2006;20(2):187–210.

149. Hanley MJ, Cancalon P, Widmer WW, Greenblatt DJ. The effect of grape-fruit juice on drug disposition. *Expert opinion on drug metabolism & toxicology.* Mar 2011;7(3):267–286.

150. Shimada T, Tanaka K, Takenaka S, *et al.* Structure-function relationships of inhibition of human cytochromes P450 1A1, 1A2, 1B1, 2C9, and 3A4 by 33 flavonoid derivatives. *Chemical research in toxicology.* Dec 20, 2010; 23(12):1921–1935.

151. Choi JS, Jo BW, Kim YC. Enhanced paclitaxel bioavailability after oral administration of paclitaxel or prodrug to rats pretreated with quercetin. *European journal of pharmaceutics and biopharmaceutics: Official journal of Arbeitsgemeinschaft fur Pharmazeutische Verfahrenstechnik e.V.* Mar 2004;57(2):313–318.

152. Duan KM, Wang SY, Ouyang W, Mao YM, Yang LJ. Effect of quercetin on CYP3A activity in Chinese healthy participants. *Journal of clinical pharmacology.* Jun 2012;52(6):940–946.

153. Li Y, Revalde JL, Reid G, Paxton JW. Interactions of dietary phytochemicals with ABC transporters: Possible implications for drug disposition and multi-drug resistance in cancer. *Drug metabolism reviews.* Nov 2010;42(4):590–611.

154. Tan KW, Li Y, Paxton JW, Birch NP, Scheepens A. Identification of novel dietary phytochemicals inhibiting the efflux transporter breast cancer resistance protein (BCRP/ABCG2). *Food chemistry.* Jun 15, 2013;138(4):2267–2274.

155. Tolson AH, Wang H. Regulation of drug-metabolizing enzymes by xenobiotic receptors: PXR and CAR. *Advanced drug delivery reviews.* Oct 30, 2010;62(13):1238–1249.

156. Dou W, Zhang J, Zhang E, *et al.* Chrysin ameliorates chemically induced colitis in the mouse through modulation of a PXR/NF-kappaB signaling pathway. *The Journal of pharmacology and experimental therapeutics.* Jun 2013;345(3):473–482.

157. Satsu H, Hiura Y, Mochizuki K, Hamada M, Shimizu M. Activation of pregnane X receptor and induction of MDR1 by dietary phytochemicals. *J Agric Food Chem.* Jul 9, 2008;56(13):5366–5373.

158. Yao R, Yasuoka A, Kamei A, *et al.* Dietary flavonoids activate the constitutive androstane receptor (CAR). *J Agric Food Chem.* Feb 24, 2010;58(4):2168–2173.

159. Yao R, Yasuoka A, Kamei A, *et al.* Polyphenols in alcoholic beverages activating constitutive androstane receptor CAR. *Bioscience, biotechnology, and biochemistry.* 2011;75(8):1635–1637.

160. Hoek-van den Hil EF, van Schothorst EM, van der Stelt I, *et al.* Quercetin decreases high-fat diet induced body weight gain and accumulation of hepatic and circulating lipids in mice. *Genes & nutrition.* Sep 2014; 9(5):418.

161. Niture SK, Khatri R, Jaiswal AK. Regulation of Nrf2-an update. *Free radical biology & medicine.* Jan 2014;66:36–44.

162. Leinonen HM, Kansanen E, Polonen P, Heinaniemi M, Levonen AL. Role of the Keap1-Nrf2 pathway in cancer. *Advances in cancer research.* 2014; 122:281–320.

163. Suzuki T, Motohashi H, Yamamoto M. Toward clinical application of the Keap1-Nrf2 pathway. *Trends in pharmacological sciences.* Jun 2013; 34(6):340–346.

164. Yang JH, Shin BY, Han JY, *et al.* Isorhamnetin protects against oxidative stress by activating Nrf2 and inducing the expression of its target genes. *Toxicol Appl Pharmacol.* Jan 15, 2014;274(2):293–301.

165. Zhai X, Lin M, Zhang F, *et al.* Dietary flavonoid genistein induces Nrf2 and phase II detoxification gene expression via ERKs and PKC pathways and protects against oxidative stress in Caco-2 cells. *Mol Nutr Food Res.* Feb 2013;57(2):249–259.

166. Sun GB, Sun X, Wang M, *et al.* Oxidative stress suppression by luteolin-induced heme oxygenase-1 expression. *Toxicol Appl Pharmacol.* Dec 1, 2012;265(2):229–240.

167. Murray IA, Patterson AD, Perdew GH. Aryl hydrocarbon receptor ligands in cancer: Friend and foe. *Nature reviews. Cancer.* Nov 24, 2014;14(12): 801–814.

168. Puppala D, Gairola CG, Swanson HI. Identification of kaempferol as an inhibitor of cigarette smoke-induced activation of the aryl hydrocarbon receptor and cell transformation. *Carcinogenesis.* Mar 2007;28(3):639–647.

169. Zhang S, Qin C, Safe SH. Flavonoids as aryl hydrocarbon receptor agonists/ antagonists: Effects of structure and cell context. *Environmental health perspectives.* Dec 2003;111(16):1877–1882.

170. Mohammadi-Bardbori A, Bengtsson J, Rannug U, Rannug A, Wincent E. Quercetin, resveratrol, and curcumin are indirect activators of the aryl hydrocarbon receptor (AHR). *Chemical research in toxicology.* Sep 17, 2012;25(9):1878–1884.

171. Kawajiri K, Kobayashi Y, Ohtake F, *et al.* Aryl hydrocarbon receptor suppresses intestinal carcinogenesis in ApcMin/+ mice with natural ligands. *Proceedings of the National Academy of Sciences of the United States of America.* Aug 11, 2009;106(32):13481–13486.

172. Puppala D, Lee H, Kim KB, Swanson HI. Development of an aryl hydrocarbon receptor antagonist using the proteolysis-targeting chimeric molecules approach: A potential tool for chemoprevention. *Molecular pharmacology.* Apr 2008;73(4):1064–1071.

173. Swanson HI, Choi EY, Helton WB, Gairola CG, Valentino J. Impact of apigenin and kaempferol on human head and neck squamous cell carcinoma. *Oral surgery, oral medicine, oral pathology and oral radiology.* Feb 2014;117(2):214–220.

174. Guarrera S, Sacerdote C, Fiorini L, *et al.* Expression of DNA repair and metabolic genes in response to a flavonoid-rich diet. *The British journal of nutrition.* Sep 2007;98(3):525–533.

Flavonoids and the Inflammatory Response*   **3**

Inflammation is often described as the response of our immune system to tissue injury and potentially harmful substances such as microorganisms, chemicals, drugs, and other materials recognized as foreign or "xenobiotic."[1] The immune response is composed of the innate and adaptive responses. The innate response is present at birth and involves basophils, neutrophils, eosinophils, mast cells, macrophages, dendritic cells, and natural killer (NK) cells. The adaptive or acquired response involves B and T lymphocytes, increases with age, and has specificity as well as

---

*** Abbreviations:** AHR, aryl hydrocarbon receptor; AKT, V-akt murine tymoma viral oncogene homolog 1; AP1, activator protein 1; ATF, activating transcription factor; BCL2, B-cell lymphoma 2; C/EBP, CCAAT/enhancer-binding protein; CD, Crohn's disease; COX-2, cyclooxygenase 2; DAMP, damage-associated molecular pattern; DSS, dextran sodium sulfate; EGCG, (-) epigallocatechin-3-gallate; FasL, Fas ligand; GM-CSF, granulocyte-macrophage colony-stimulating factor; HIF1/2/3α, hypoxia inducible factor 1/2/3 alpha; IFNγ, interferon gamma; IGF1, insulin growth factor 1; IL1β/IL6/IL8/IL10/IL12, interleukins 1 beta/6/8/10/12; IBD, inflammatory bowel disease; IKKβ, inhibitor of nuclear factor kappa-B kinase subunit beta; iNOS, inducible nitric oxide synthase; JUN, jun protooncogene; M1 macrophages, classically activated macrophages; M2 macrophages, alternatively activated macrophages; MAF, V-maf musculoaponeurotic fibrosarcoma oncogene; MAPK, mitogen-activated protein kinase; MYC, v-myc avian myelocyomatosis viral oncogene homolog; NFAT, nuclear factor of activated T cells; NFκB, nuclear factor of kappa light polypeptide gene enhancer in B-cells; NK, natural killer cells; NOTCH, neurogenic locus homolog; NRF2, nuclear factor, erythroid 2-like 2; PAMP, pathogen-associated molecular pattern; PPAR, peroxisome proliferator activated protein; PXR, pregnane X receptor; RANKL, receptor activator of nuclear factor kappa-B; RELB, v-rel avian reticuloendotheliosis viral oncogene homolog B; RONS, reactive oxygen and reactive nitrogen species; STAT3, signal transducer and activator of transcription 3; TAM, tumor-associated macrophages; TGFβ, transforming growth factor beta; TLR, toll-like receptor; TNBS, trinitrobenzene sulfonic acid; TNFα, tumor necrosis factor alpha; TNFAIP3, tumor necrosis factor, alpha-induced protein 3; TP53, tumor protein 53; UC, ulcerative colitis.

memory. The primary signals transmitted and received during an immune response are mediated by cytokines and chemokines and involve a variety of innate, adaptive, epithelial, and muscle cells as well as fibroblasts. Cytokines are defined as small peptides or glycoproteins whereas chemokines are a subclass of cytokines that function as chemotactic molecules. The effects of cytokines are mediated by specific receptors that include interferon receptors, and the TNF receptor family. Appropriate signaling by chemokines and cytokines is required for the development of an organism, maintenance of tissue homeostasis, and activation and function of the immune system. However, inappropriate or excessive production of cytokines and chemokines often contributes to the pathogenesis of chronic disease states such as diabetes, inflammatory bowel diseases, and cancer. Thus, cytokines and chemokines are often dichotomous in their functions. They play both protective and pro-inflammatory roles depending on their cellular source, the absence or presence of other environmental cues, and whether their expression is transient or sustained.[2]

The innate immune response is often described as the first line of defense in the recognition of microorganisms or tissue injury.[3] The innate immune response of mucosal tissue, for example, involves macrophages, dendritic cells, and epithelial cells that reside at the surface and respond in an immediate, early and non-specific fashion. Damage to the epithelial layer can arise from mechanical lacerations or from the presence of allergens with proteolytic activity (e.g., phospholipase A present in bee venom) or infectious agents. The link between the innate and adaptive immune responses is provided by the professional antigen-presenting cells, in particular, macrophages and dendritic cells. Upon activation, macrophages and dendritic cells guide the differentiation of the naïve CD4 cells into four subsets of T-helper cells, the T_H1, T_H2, T_H17, and T_{reg}. These subsets of cells are characterized by the predominate cytokines that they secrete. T_H1 cells are best known for their production of IFNγ as well as TNFα. The T_H2 cells secrete primarily IL4, IL5, and IL13, whereas the T_H17 cells produce IL17, IL21, and IL22. Finally, the T_{reg} cells secrete the anti-inflammatory cytokine IL10. The T_H17-mediated response plays a critical role in balancing the anti-inflammatory versus pro-inflammatory responses as its primary function is to restrain the T_{reg} cells from suppressing the T_H1 response. With respect to inflammation-associated tumorigenesis, the T_H2

and T_H17-mediated responses are the most strongly implicated. For example, a T_H2 signature is associated with poor prognosis of pancreatic, hepatocellular, non-small lung carcinoma, gastric, and breast cancers.[4] Similarly, a T_H17-dominated signature is often observed in patients with colorectal tumors who are diagnosed with a very poor prognosis.[5]

Macrophages, in addition to effector T cells, also play important roles in tumorigenesis.[6,7] Monocytes are recruited to the specific site of tissue injury or infection where they differentiate into macrophages. Within the homeostatic tissue, macrophages play a protective role and exert primarily anti-inflammatory activities. In the presence of bacteria, protozoa, and viruses, monocytes are "classically activated" into a subset referred to as M1 macrophage. In addition to their role in host defense, M1 macrophages actively participate in anti-tumor activities. The M1 macrophages typically secrete high levels of $TNF\alpha$, IL12, and RONS. The "alternatively activated" subset of macrophages known as M2 macrophages exert anti-inflammatory functions that play key roles in appropriate wound healing. Their activation arises in response to the presence of IL4, IL13, IL10, and the glucocorticoid hormones. A third subset, termed "regulatory macrophages", are capable of secreting high levels of the anti-inflammatory cytokine IL10. Finally, a subset of macrophages, called tumor-associated macrophages (TAM), reside within the microenvironment. TAM are often referred to as "M2-like" as their phenotype, in particular their ability to promote angiogenesis and tissue remodeling, most closely resembles that of the alternatively activated macrophages. The accumulation of TAM plays an important role in tumor growth and the ability of the tumor to evade the actions of cytotoxic T cell responses. Thus, M1 macrophages are anti-tumorigenic while M2, regulatory, and TAM are pro-tumorigenic. However, during chronic inflammation, the abundant M1 macrophages may convert to an M2-like phenotype as the initiated tumors begin to invade the surrounding tissue and vascularize. Interestingly, blocking the activities of NFκB can switch TAM to a phenotype that is more M1-like and anti-tumorigenic. Macrophages are thought to play an as yet poorly understood role in the resolution of inflammation and restoration of tissue homeostasis.

As mentioned above, activation of innate immunity involves the sensing of the presence of pathogens by macrophages and dendritic cells.[8]

This sensing is facilitated by the recognition of common features of the pathogens by specific pattern recognition receptors (often, the toll-like receptors) expressed by resident macrophages and immature dendritic cells. For example, lipopolysaccharides on gram-negative bacteria are recognized by toll-like receptor 4 (TLR4) whereas the double stranded DNA of viruses is recognized by toll-like receptor 9 (TLR9). Pathogen-sensing is initiated by the presence of PAMPs, pathogen-associated molecular patterns. However, sterile inflammation, like that which occurs during cancer, is initiated by the presence of endogenous molecules termed DAMPs or damage-associated molecular patterns. Both PAMPs and DAMPs bind and activate the same pattern recognition receptors, in particular, TLR2 and TLR4, and thereby trigger similar immune responses. DAMPs arise from several different sources. They are generated as fragments of proteolytic processes that occur within the extracellular matrix. Both cytokine-activated macrophages and tissue resident cells are capable of initiating de novo synthesis of DAMPs from the extracellular matrix. They are also present as proteoglycans that are released from the activities of enzymes such as metallomatrix proteinases 2, 3, and 13. In addition, uric acid, the ultimate product formed during purine nucleotide catabolism, can act as a DAMP. Finally, cells dying from necrosis and apoptosis or undergoing autophagy release DAMP-acting molecules. This latter source of DAMPs is indicative of the intricate relationship that is emerging between the pathways that mediate pathogen recognition, inflammation, and cell death.

The link between chronic inflammation and tumorigenesis. The inflammatory response is a normal, physiological response to infections or following exposures to pathogens.[9,10] In these scenarios, the inflammatory response is self-limiting and is followed by its resolution which is mediated by the downregulation of pro-inflammatory cytokines (i.e., TNFα, IL1β, and IL6) and upregulation of anti-inflammatory cytokines, such as IL10. However, if resolution is not achieved, the pro-inflammatory events are chronically active and play a pathogenic role. Chronic inflammation is the common denominator of a plethora of cancer risk factors including chronic infections, inhaled pollutants, obesity, and autoimmunity, and is indicative of the critical role that inflammation plays in tumor development. In addition, pathologies of chronic inflammation, such as inflammatory bowel diseases, often lead to the development of cancer (i.e., colorectal cancer). Chronic inflammation impinges on the tumorigenic process at the initial steps of tumor formation and throughout the development, growth, and progression of the metastatic tumor. Its effects are intrinsic and extrinsic, local and systemic. At the intrinsic level, the signaling cascades initiated by the activated oncogenes participate in the recruitment of inflammatory cells. At the extrinsic level, the conditions of chronic inflammation, which includes the generation of reactive oxygen species, contribute to the formation of the neoplastic cell.

As an initiator of tumor formation, chronic inflammation damages DNA and contributes toward telomere shortening. Here, the overzealous inflammatory response accelerates the formation of reactive oxygen and reactive nitrogen species (referred to collectively as RONS) that damage DNA and interfere with the DNA damage response, thus aiding in genomic instability. During tumor progression, this abundance of RONS drives proliferation of the cell harboring the mutated DNA sequences and enhances the ability of this cell to migrate, invade the circulatory system, and survive the "hostile" environment of the new host tissue. In addition, the inflammatory response contributes to the formation of new blood vessels that surround the tumor mass. The majority of solid tumors are composed of a heterogeneous group of cells that includes the neoplastic epithelial cell, cancer stem cells, endothelial cells, cancer associated fibroblasts, and immune inflammatory cells. Within this "local" environment, the immune inflammatory cells can exert either pro- or anti-tumor signals depending on cues elicited by their neighboring cells and within their own cellular milieu. Increasing evidence indicates that signaling from within the tumor tissue also elicits pro-tumorigenic responses from the host at distant tissue sites and in this manner, comprises a systemic inflammatory response.[11] In addition to its role in mediating tumor progression and metastasis, systemic inflammation is thought to underlie many symptoms experienced by the cancer patient such as weight loss, fatigue, pain, and depression.[12]

Role of cytokines in tumor-promoting inflammation. The cell-to-cell communication that occurs within the tumor microenvironment and between the normal cell and tumor cell is mediated by cytokines.[7,13] The cellular sources of cytokines implicated in this crosstalk include immune cells, fibroblasts, endothelial cells, and some epithelial cells. They regulate proliferation, survival, differentiation, immune cell activation, and migration, and can induce either anti-tumor (by initiating an anti-tumor response) or pro-tumor (by increasing cell transformation and malignancy) events depending on other conditions within their microenvironment. For example, within a few hours following their activation, the macrophages begin to secrete IL23. In response, the resident macrophages produce IL17 which encourages the stromal, epithelial, and endothelial cells, as well as some monocytes, to produce IL1, IL6, IL8, the CXC ligand 1 and TNFα. These cytokines serve to recruit neutrophils to the site of injury or infection which upon activation, release IL1 and IL8 and synthesize serine proteinases such as myeloperoxidase as well as metalloproteinases. Inflammation is resolved when neutrophils have completed their phagocytic tasks and switch to a more anti-inflammatory pro-resolution state. In addition, by undergoing apoptosis, the neutrophils simulate the macrophages to also assume a pro-resolution phenotype. It is thought that cytokines produced by the host tissue, such as IL1, IL8, and granulocyte-macrophage colony-stimulating factor (GM-CSF) contribute to neutrophil survival and the accumulation of neutrophils that are not eliminated by macrophages during persistent inflammatory conditions.

The most widely implicated pro-inflammatory and pro-tumorigenic cytokines are TNFα and IL6 which are often highly expressed during chronic inflammation. The pro-tumorigenic effects of TNFα, IL6, and TGFβ are primarily due to increased levels of reactive oxygen and nitrogen species which induce cellular oxidative damage; damage lipids, proteins, and DNA; and upregulate the activities of key transcription factors. High levels of IL6 are often detected in the serum of patients diagnosed with systemic cancer. IL6 plays a major role in the promotion of tumor cell proliferation while also inhibiting apoptosis. The conversion of normal cells to tumor cells is promoted by both IL6 and TNFα.

Cytokines with anti-inflammatory activities, in particular TGFβ and IL10, can exert either pro- or anti-inflammatory effects depending on the cell type and stage of the tumorigenic process. During early tumorigenesis, TGFβ acts as a tumor suppressor primarily by inhibiting cell cycle progression and inducing apoptosis. At later stages, however, increased levels of TGFβ aid the tumor cell in its ability to undergo epithelial to mesenchymal transition, invade surrounding tissue, and successfully metastasize to distant tissues. Within the tumor microenvironment, TGFβ is synthesized by the tumor cell and a variety of immune cells and fibroblasts. IL10 is expressed in all immune cells and a restricted number of tumor cells. In many cancer patients, the high serum levels of IL10 are thought to be due to its expression by tumor-infiltrating suppressor macrophages. A key regulator of IL10 expression levels is the transcription factor STAT3. IL10 in turn, rapidly activates STAT3 in a sustained manner. The anti-inflammatory and anti-tumor activities of IL10 arise primarily from its inhibition of NFκB and its interaction with and direct inhibition of a number of cytokines. The pro-tumor activity of IL10 is thought to be a consequence of its interference with anti-cancer immunity.

Role of transcription factors in tumor relevant inflammation. At the molecular level, the key players that mediate the crosstalk between chronic inflammation and carcinogenesis are redox-sensitive transcription factors such as NFκB,[9] AP1,[14,15] STAT3,[16] HIF1α (hypoxia inducible factor 1α)[17] and NRF2 (nuclear factor, erythroid 2-like 2).[18] These transcription factors regulate and can in turn be regulated by cytokines (i.e., IL6, IL10, TNFα, and TGFβ) and enzymes (i.e., cyclooxygenase 2, COX-2 and iNOS). Typically, redox-sensitive transcription factors are activated following exposure to oxidative and/or pro-inflammatory stimuli. They activate a unique set of genes and/or coordinately activate a battery of genes. These transcription factors are often constitutively activated within the tumor tissue by the consistent presence of oxidative stress and high expression of pro-inflammatory conditions.

The NFκB pathway is upregulated by cytokines and pathogenic molecules via activation of their cognate receptors (e.g., TNF and toll-like receptors).[9] Two distinct NFκB pathways can be activated: the canonical and the noncanonical. The canonical pathway is stimulated by interleukins, is associated with innate immunity, and recruits immune cells such as neutrophils. Activation of the canonical pathway involves phosphorylation and nuclear translocation of p65 as well as formation of the DNA binding complex composed of p65/p52. The noncanonical pathway is associated with adaptive immunity such as that stimulated by RANKL and is mediated by formation of the RELB/p52 DNA binding dimer. Negative regulation of the active NFκB pathway is accomplished by several independent pathways. Induction of IL10 by activated STAT3 reduces the cellular levels of key stimulators of NFκB, in particular, TNFα, IL6, and IL12. Other regulators such as TNFAIP3 (tumor necrosis factor, alpha-induced protein 3) initiate stepwise ubiquitination, and proteosomal degradation is initiated by proteins such as A20. Constitutive activation of NFκB that occurs during chronic inflammation is thought to arise from either an ineffective anti-inflammatory response (which may be due to deregulation of the NFκB pathway) or persistent pathogenic infection. It contributes to pro-tumorigenic activity that includes increased expression of anti-apoptotic genes, cytokines that regulate the immune response (TNFα, IL1, IL6, and IL8), adhesion molecules that aid in the recruitment of leukocytes, and gene products that facilitate epithelial to mesenchymal transition and tumor vascularization. The pro-survival/pro-proliferative activity of constitutive NFκB is thought to involve its antagonism of the tumor suppressor TP53. In addition, NFκB's activities are enhanced by AKT. As will be discussed below, nuclear receptors such as PXR, AHR, and NRF2 participate in mutual suppression of NFκB.

The AP1 (activator protein 1) family of transcription factors is composed of three subfamilies — JUN, ATF, and MAF — that harbor a common basic leucine zipper domain that is essential for their dimerization and DNA binding activities.[14,15] The AP1 family regulates cell fate decisions (proliferation, differentiation, and apoptosis), cell migration and invasion, inflammation, and angiogenesis. Like NFκB, AP1 is activated by oxidative and pro-inflammatory stimuli; however, its activation by the MAPK signaling pathways is thought to play a more important role. Depending on the stimuli as well as its duration, AP1 can exert pro-proliferative (via upregulation of cyclin D) or anti-proliferative (via suppression of p53 and induction of anti-apoptosis genes) activities. In addition, activation of AP1 can upregulate genes such as Fas ligand (FasL) resulting in pro-apoptotic events. Upregulation of AP1 can also enhance expression of COX-2, iNOS, and cytokines such as IL1β and TNFα. Crosstalk between AP1 and NFκB occurs via their mutual upregulation by pro-inflammatory cytokines (i.e., TNFα, IL1β, IL6, and IL8) and their coordinate upregulation of genes involved in inflammation (i.e., COX-2), matrix remodeling, and proliferation.

Signal transducer and activator of transcription 3 (STAT3) is also activated by cytokines, in particular IL6 and IL10, as well as a wide variety of agonist-induced growth factor receptors (epidermal growth factor receptors, fibroblast growth factor receptors, insulin-like growth factor receptors, hepatocyte growth factor receptors, platelet-derived growth factor receptors, and vascular endothelial growth factor receptors).[16] More than 40 polypeptides have been shown to initiate STAT3 phosphorylation and activation. STAT3 is often described as being poised at the convergence of the many signaling pathways that are triggered by cytokines, growth factors, and oncogenes. As such, its activation results in the upregulation of genes involved in proliferation, cell survival, inflammation, invasion, metastasis, and angiogenesis. In addition, STAT3 acts coordinately with NFκB to alter gene regulation. STAT3 expression is linked to inflammatory processes associated with gastric, colon, liver, lung, and pancreatic cancers. For example, in colitis-associated colorectal cancer, IL6 activation of STAT3 plays a key role in the early stages of tumorigenesis. STAT3-upregulation of the expression of key cell cycle proteins (Cyclin D1, Cyclin D2, Cyclin B, and MYC) underlies its pro-proliferation activities. STAT3-upregulation of the expression levels of anti-apoptotic proteins (BCL2 and BCL2-like 1) facilitates its pro-survival activities. The participation of STAT3 in the NFκB-IL6-STAT3 signaling cascade that occurs during oncogene activation is thought to play a key role in both tumor initiation and progression.

The hypoxia inducible factors (HIF) are typically upregulated during conditions of oxygen deprivation or hypoxia.[17] As a transcriptionally active complex, HIF is composed of two subunits. The oxygen-sensitive subunits are HIF1α, HIF2α, and HIF 3α whereas the oxygen-insensitive subunit is HIF1β (also referred to as ARNT, aryl hydrocarbon receptor nuclear translocator). However, HIF1α is also activated by cytokines (e.g., TNFα and IL1β), hormones such as insulin like growth factor 1 (IGF1), and vasoactive peptides. The nuclear localization and hence, the ability of HIFα to function as a transcriptional activator, is regulated by HIF prolyl hydroxylase domain enzymes. HIF target genes are involved in angiogenesis, glycolysis and erythropoiesis, proliferation and metastasis.

Chronically inflamed tissues contain regions of hypoxia which are thought to arise from the disruption of the microvasculature and increased oxygen consumption by the inflamed resident cells and the infiltrating, adaptive immune cells. In addition to hypoxia, HIF is upregulated by IL1β via NFκB and contributes to the inflammatory response by altering the function, differentiation, and survival of immune cells. Within the context of the tumor microenvironment, HIFα appears to play dual roles in inducing formation of either the pro-inflammatory or anti-inflammatory forms of key immune cells (macrophages and T cells).

Oxidative and electrophilic stresses activate the NRF2 transcription factor which in turn induces a cytoprotective response involving the upregulation of a battery of gene products capable of generating antioxidants, harboring antioxidant activities, repairing damaged DNA, exporting toxic substances, and inhibiting the activities of pro-inflammatory cytokines.[19] NRF2 often accomplishes this myriad of events by collaborating with other transcription factors such as the AHR, NFκB, TP53, and NOTCH. Since numerous pre-clinical studies have shown that NRF2 activators are effective chemo preventive agents, several clinical trials are currently ongoing to test their effectiveness in human subjects. Although these studies are as yet incomplete, there is some evidence that depending on the agent under consideration, significant side effects may occur. Further, it is becoming increasingly clear that NRF2 plays a dual role in the carcinogenic process. While it provides the normal cell with a tool to survive a stressful environment, this same adaptive response can be exploited by a tumor cell to gain a growth advantage.

Other transcription factors that have been implicated in cancer-promoting, chronic inflammation include the AHR, PXR, and PPARs. The AHR is a regulator of proliferation and differentiation of T cells, B cells, and dendritic cells, as well as the expression levels of pro-inflammatory cytokines such as IL6, IL12, and TNFα.[20,21] The AHR likely contributes to the chronic inflammatory conditions within the tumor microenvironment via both its direct regulation of cytokine and chemokine expression as well as its crosstalk with NFκB. The role of PXR in inflammation has been best described within the context of inflammatory bowel diseases, but is also implicated in colon cancer and chronic inflammation and tumorigenesis of the esophagus.[22] The anti-inflammatory events associated with agonist-induced activation of PXR appears to involve its repression of NFκB and subsequent downregulation of pro-inflammatory cytokine expression. Its crosstalk with NFκB is mutual as lack of PXR expression results in enhanced expression of pro-inflammatory cytokines in the gut. The PPARs are also thought to play important roles in modulating inflammation, particularly that associated with obesity and inflammatory bowel diseases.[23] Agonists of PPARα, PPARβ/δ, and PPARγ exhibit anti-inflammatory properties. While the anti-inflammatory mechanisms that are elicited following exposures to PPAR agonist have yet to be fully elucidated, it appears that both transactivation and transrepression of genes are involved. For example, transrepression of PPARα is mediated by its interference of STAT, AP1, and/or NFκB activities. In addition, PPARγ physically interacts with the NFκB transcriptional complex thereby repressing its transcriptional activation. PPAR-induced genes include the interleukin 1 receptor antagonist which has been shown to be upregulated by agonists of PPARα.

Impact of flavonoids on inflammation. Flavonoids that have been shown to be effective at inhibiting the pro-inflammatory response include apigenin,[24] chrysin,[25] isoliquiritigenin,[26] isorhamnetin,[27] kaempferol,[28] luteolin,[29] naringenin,[30] and quercetin.[31,32] A common mechanism by which many of these flavonoids (i.e., apigenin, chrysin, luteolin, isorhamnetin, quercetin, and kaempferol) appear to exert their anti-inflammatory effects is via inhibition of NFκB. In-depth study of apigenin, luteolin, genistein, and 3′-hydroxy-flavone revealed that these flavonoids can inhibit a number of events that lead to NFκB gene activation, including the activity of IKKβ, and DNA binding and transcriptional activation of the NFκB heterodimers, p65/p52 and/or Rel/p52.[33] The most plausible underlying common mechanism by which flavonoids inhibit NFkB is via their inhibition of kinases that are required for NFκB activation, such as AKT or MAPK.[34,35]

A number of flavonoids have been shown to exert their anti-inflammatory properties via activation of several ligand activated transcription factors, such as NRF2, PXR, PPAR, and AHR, as well as repression of NFκB. For example, isorhamnetin,[36] genistein,[37] and luteolin[38] upregulate NRF2 which is capable of participating in a NFκB regulatory feedback loop.[39]

Similarly, the ability of chrysin to inhibit intestinal inflammation has been shown to involve its activation of PXR and engagement of the mutually repressive interaction between PXR and NFκB.[25] Finally, the activation of PPARγ and subsequent anti-inflammatory activities induced by flavonoids such as baicalin,[40] chrysin,[41] hesperidin,[42] and wogonin[43] appear to involve PPARγ-mediated repression of NFκB. Naringenin has also been shown to act as an AHR agonist and at least *in vitro*, to exert immunosuppressive effects.[44] However, the extent to which these flavonoids impact inflammation via their modulation of AHR activity *in vivo* in addition to the underlying mechanisms is yet to be determined.

Additional mechanisms by which flavonoids have been proposed to exert their anti-inflammatory effects include inhibition of STAT3 (i.e., morin and quercetin),[32,45] inhibition of C/EBP beta (i.e., baicalin),[46] and inhibition of nitric oxide synthesis (i.e., kaempferol and naringenin).[30,47] Currently, it does not appear likely that the anti-inflammatory effects of flavonoids arise from either their inhibition or activation of a single pathway. More likely is that flavonoids impinge on an interrelated network of kinases and transcription factors that can coordinately exert a protective response against the pathological stress imposed by chronic inflammation.

Evidence that the chemopreventive properties of flavonoids involve anti-inflammation. In order to determine whether the chemopreventive effects of flavonoids are mediated by their anti-inflammatory activities in the human population, appropriate biomarkers of chronic inflammation must be used. Markers of inflammation that have been shown to be associated with cancer risk and cancer progression and/or survival include systemic cytokine concentrations (i.e., serum levels of IL6 and IL8), measures of leukocyte quantities, expression of acute phase proteins such as C-reactive protein and oxidative DNA damage.[48] A recently published review of chronic human intervention studies has examined the impact of flavonoids on either inflammatory cytokines in serum levels of inflammatory cytokines or *ex vivo* markers of immune function.[49] Here, the interventions consisted of either consumption of flavonoid-rich foods such as fruit juice, grape extracts, or wine or administration of pure quercetin. The most commonly measured markers of inflammation were serum levels of IL6 and TNFα. These authors report that *in vivo* supplementation of flavonoid-rich foods or quercetin did not significantly affect measures of inflammation. Interestingly, even administration of pure quercetin (150–1,000 mg) that resulted in serum levels of quercetin as high as 1,000 µg/L failed to consistently reduce the serum levels of IL6 or TNFα. One caveat noted by the authors is that the health status of the human subjects prior to the intervention may be of importance. That is, those with significantly higher plasma levels of inflammatory cytokines may prove to be more responsive to the effects of the supplemented flavonoids.

Other studies of interest are those in which consumption of flavonoids have been linked to changes in serum cytokine levels in patients with diagnosed tumors. Bobe *et al.* reported on a four-year randomized nutritional intervention trial in participants diagnosed with colorectal adenomas.[50] A total of 1905 participants were involved with 958 in the nutritional intervention arm. Within the first three year period, the flavonoid consumption of the group in the nutritional intervention arm had increased two-fold. At the end of four years, 40% of the participants had at least one adenoma. Serum levels of eight different cytokines (IL1β, IL2, IL8, IL10, IL12p70, GM-CSF, IFNγ and TNFα) were examined in relation to flavonal intake and colorectal adenoma recurrence. While there was no association between serum cytokine profile and colorectal adenoma recurrence, a decrease in cytokine concentration in individuals with high flavonol consumption (>29.7 mg/day) was observed.

A similar study was performed to determine whether high intake of freeze dried black raspberries would be associated with changes in plasma cytokine levels and beneficial changes in colorectal tissue.[51] In this pilot study, six female and 18 male subjects with adenocarcinomas were administered 20 grams of black raspberry powder three times per day for three weeks (until surgical resection of the adenocarcarcinoma). Tissue biopsies and serum samples were obtained prior to and following the intervention. The tissue samples were evaluated for markers of proliferation, angiogenesis, and apoptosis and the plasma samples were monitored for expression levels of cytokines. The results from this study indicated that the berry extracts increased plasma levels of GM-CSF and decreased plasma levels of IL8. Further, these changes in plasma cytokine levels correlated with beneficial changes of the berry extracts observed in the colorectal tissues. Thus, this study, together with that of Bobe *et al.*,[50] indicates that certain cytokines may serve as biomarkers to be used for evaluating the anti-inflammatory and chemopreventive effects of flavonoids.

Inflammatory bowel disease (IBD), a disease of chronic inflammation. The link between chronic inflammation and tumorigenesis may be best examined using IBD as a specific example, given that it ranks amongst the highest risk factors for developing colorectal cancer.[52] IBD occurs as either Crohn's disease (CD) or ulcerative colitis (UC) with a peak onset of 15–30 years of age.[53,54] Common features of IBD are conditions of chronic relapse and remitting intestinal inflammation. CD can affect any section of the gastrointestinal tract and typically occurs as mucosal lesions that are randomly distributed. In contrast, UC is limited to the colon, initiating at the rectum and progressing into the colon with an increase in disease severity.[55] The incidence of IBD is increasing in many regions of the world, and IBD is considered an emerging global disease.[56] Countries with the highest incidences and prevalence are typically those that are more developed and include Canada, Europe, and the United States. For example, in Canada, the incidence rate of UC is 19.2/100,000 whereas that of CD is 20.2/100,000. While the incidence rates of IBD in Asia is increasing, it is considerably lower ranging from 0.1–6.3/100,000 (UC) and 0.04–5.0/100,000 (CD). IBD is thought to arise from a combination of factors, including genetics, a dysregulated immune response, and environmental triggers. Genetic studies indicate that gene products involved in the elimination of endogenous small organic cations, drugs, and environmental toxins are linked to IBD etiology. A link between environmental exposures and CD is supported by the fact that 40–50% of individuals with identical genetic makeup are discordant for CD. Further, the worldwide increase in incidence and prevalence of IBD is associated with a westernized lifestyle which includes changes in the diet, intestinal colonization, and increased industrialization.[57] A well-characterized environmental risk factor is tobacco smoke; patients with familial CD who smoke are reportedly at an enhanced (two-fold) risk of developing CD.[58] However, findings of both negative[59] and positive[59,60] correlations between other environmental exposures and either diagnoses or severity of IBD have been reported.

Pathology of IBD: immune dysfunction and role of T-helper cells. The intestine harbors the largest surface contact between the body and external environment and is thus equipped with the largest lymphoid organ of the body.[61] Intestinal homeostasis requires a dynamic interplay between the immune system, intestinal tissue, intestinal flora, and dietary compounds present within the lumen. Chronic intestinal inflammation arises in genetically susceptible individuals when breakdown in the regulatory processes occur.[62] As a consequence, the immune response to the luminal content becomes exaggerated and inappropriately pro-inflammatory. Both the innate and adaptive immune responses are thought to be involved.

The GI tract is a site of exposure to both deleterious pathogens and probiotic bacteria.[63–65] Given this, members of the innate immune system residing in the gut mucosa are activated by PAMPs. The default state of the gut is one of hyporesponsiveness where its response to pathogens is attenuated and the presence of commensal bacteria and food antigens is tolerated. This state of oral tolerance is formally defined as the suppression of immune responses to antigens that have been administered previously by the oral route. Patients with IBD lack this tolerance and instead present with a hyperactive response to the luminal contents. Defects in the epithelial barrier can also lead to the development of IBD. While the normal epithelium of the intestinal mucosa typically forms a barrier that excludes luminal substances, in patients with IBD, this epithelial barrier becomes more permeable. Involved genes associated with IBD include mutated forms of the organic cation transporter which is thought to contribute to impaired fatty acid oxidation. Defects in the epithelial barrier also contribute to a lack of anti-microbial production, impaired microbial sensing, and defective microbial clearance. The epithelial cells react to the presence of excess microorganisms by enhancing their secretion of pro-inflammatory cytokines and thereby contribute to the pro-inflammatory status of the mucosal tissue.

Macrophages and dendritic cells are also involved in the pathogenesis of IBD. Due in part to impaired TLR sensing and signaling, their pathogenic responses are enhanced while their protective responses are inadequate. The macrophages from patients with IBD display an impaired ability to clear bacteria leading to bacterial persistence and chronic stimulation. This is thought to involve their production of more pro-inflammatory cytokines (IL12 and IL23) but of less protective cytokines (IL10 and GM-CSF). In addition, the macrophage phenotype is altered and is typically composed of few M1 macrophages and an overabundance of the M2 macrophage subtype. This results in a full rather than a physiologically attenuated macrophagic response which normally exists within the homeostatic tissue.

Dendritic cells that reside in the gut mucosa directly sense the contents of the gut lumen and recognize antigens transported into the lamina propria, the tissue immediately underlying the mucosal epithelium. By using the PAMPS and pattern recognition receptors such as the TLRs, the dendritic cells are able to discriminate between pathogenic versus commensal bacteria. Depending on the accuracy and sensitivity of their sensing pathways that are relayed to the adaptive immune cells, dendritic cells can dictate whether the population of effector T cells is represented by those expressing pro-versus anti-inflammatory cytokines. As previously mentioned, activated dendritic cells play a key role in determining the composition of the T_H1, T_H2, T_H17, or T_{reg} populations that arise from the naïve CD4+ T cells.[55,64–66] However, the environmental milieu of the cytokine population also plays a role. Formation of the T_H1 subpopulation that expresses primarily IFNγ and TNFα is induced by IFNγ and IL12, whereas the T_H2 subpopulation (expressing primarily IL4, IL5, and IL13) is induced by IL4. Induction of the T_H17 subpopulation (expressing primarily IL17A, IL17F, IL21, and IL22) is mediated by both TGFβ and IL6 and their expansion is promoted by IL23. While IL17 is considered to be a pro-inflammatory cytokine, IL22 is protective during IBD and plays a key role in tissue repair.[67] Finally, formation of the T_{reg} subpopulation is induced by TGFβ and IL27. In the healthy individual, dendritic cells are hyporesponsive which appears to be due to their ability to steer the T cell differentiation process toward the generation of a T_{reg} subpopulation which expresses high levels of the anti-inflammatory cytokine IL10. While controversial, CD is thought to be primarily due to a T_H1-mediated response as these patients express high levels of gut IFNγ and TNFα as compared to that of the healthy control. UC is typically thought to be a

T_H2-mediated disease with higher levels of IL13 expressed in the gut tissue. Over-representation of T_H17 cells appears to occur in both CD and UC, given that high levels of IL17A are expressed in the tissues of the affected patients as compared to that of the healthy controls. Cytokines such as TNFα, IL1β, and IL6 are also abundantly expressed within the inflamed mucosa, but they are more promiscuous in their actions and are associated with both forms of IBD to a lesser or greater degree. They generally arise from the stimulated innate immune cells, are secondary to the primary T_H1/T_H17 or T_H2-like responses, and propagate the pro-inflammatory response. TNFα, IL1β, and IL6 activate the NFκB pathway whereas IL6 and perhaps IL1β play an essential role in inducing the T_H17 responses. Thus, the intestinal inflammation associated with IBD is thought to arise from an "overly exuberant" response from the T_H1 and T_H17 (pro-inflammatory) subpopulation of T cells and an inadequate T_{reg}-cell (anti-inflammatory) response.

Therapeutic treatment of IBD. IBD as a disease state currently has no known medical or surgical cure.[68–70] The primary goal of IBD therapeutics is to reduce inflammation such that remission is induced and maintained. The first step in initiating therapy is to assess the patient with respect to disease severity and activity and their response to previous medications. Current IBD treatments either act to inhibit inflammation or suppress the immune system. Patients diagnosed with mild forms of IBD (UC or CD) are often treated initially with 5-aminosalicylates which include sulfalazine and mesalamine. The pharmacological target of these drugs is poorly defined but is thought to act similarly to salicylate and inhibit the production of prostaglandins, the enzyme activity of cyclooxygenases, and the transcriptional activity of NFκB. Patients with UC may also be treated with topical glucocorticoids whereas patients with CD are often co-administered antibiotics or the glucocorticoid budesonide. Patients with moderate to severe forms of IBD are often treated with the glucocorticoid prednisone and immunomodulators such as methotrexate and the thiopurines, azathiopurine, and 6-meracaptopurine. Patients with the most severe forms of IBD are typically treated with additional immunomodulators such as inhibitors of calcineurin (i.e., Cyclosporine A) or TNFα (i.e., infliximub and adalimumab) as well as surgical interventions. Other drugs that may be used include those that target adhesion molecules expressed by leukocytes, suppress the actions of IL12 and IL23, and inhibit the JAK/STAT pathway and chemokines involved in leukocyte trafficking. Ongoing research is investigating the potential effectiveness of probiotics, enteral nutrition, and fecal transplantations designed to restore the gut microbiome balance. While there is considerable interest in using nutritional supplements and traditional medicines that contain flavonoids, evidence obtained from well-controlled clinical trials is not yet sufficient for recommending their use in controlling IBD.

Glucocorticoids such as budesonide and prednisone can be effective at inducing mild to moderate remission, but typically are not effective at maintaining remission. Budesonide is a glucocorticoid formulated to be active primarily in the distal ileum and right colon. It is rapidly metabolized in the liver and thus exerts limited systemic effects. Systemic glucocorticoid effects are considerably greater when prednisone is used. Glucocorticoids exert their anti-inflammatory effects primarily by binding to and activating the glucocorticoid receptor which then represses the pro-inflammatory actions of NFκB.[71] Glucocorticoids initiate their effects relatively quickly, and are easy to use and inexpensive. However, systemic exposure to glucocorticoids often leads to serious side effects such as the increased development of opportunistic infections, osteoporosis, and metabolic disturbances that exacerbate conditions such as diabetes. Further, approximately 30% of IBD patients are resistant to the remission-induced effects of glucocorticoids.

Methotrexate and the thiopurines are immune modulators that are thought to trigger a T_H2 response that dampens the T_H1 mediated response.[69,72] These drugs are often referred to as "anti-metabolites" as their metabolites are incorporated into DNA and in this manner, reduce the proliferation of T and B lymphocytes while also abrogating a number of actions involved in the immune response. As compared to the glucocorticoids, their onset is considerably slower and is often not observed until up to three months following their administration. The thiopurines can be effective in maintaining remission in both UC and CD patients. In addition, they have been observed to enhance the efficacy of anti-TNF therapies. Adverse effects of thiopurines include myelosuppression.The extent to which myelosuppression occurs is dependent on genetic polymorphisms of thiopurine S-methyl transferase, a major enzyme responsible for their catabolism. Genetic testing for these polymorphisms can be used to improve dosage formulations and limit the myelosuppressive side effects of these drugs.

Methotrexate exhibits a more rapid response than that of the thiopurines and is often used for patients with active CD.[72,73] It is also used in patients who may no longer respond to thiopurines. Finally, it is often used in combination with anti-TNFα therapies to enhance their effectiveness. The evidence supporting the effectiveness of methotrexate in inducing and maintaining remission in patients with UC is conflicting and yet to be clearly established. Adverse effects of methotrexate that may be of concern in some patients includes stomatitis, myelosuppression, and hepatotoxicity that may contribute to liver cirrhosis.

Anti-TNFα inhibitors such as infliximub and adalimumab are termed biologics as they are monoclonal antibodies that are designed to specifically recognize and bind to TNFα, thereby inhibiting its ability to activate the TNFα receptor. They have been shown to be particularly effective in corticosteroid-dependent patients with either CD or UC and in patients with CD who have fistulas. Some risks of hypersensitivity reactions, lymphomas, nonmelanomas, and infections have been reported. In addition, approximately 30% of patients fail to respond. Other disadvantages associated with the use of anti-TNFα inhibitors is their expense and their delivery which is restricted to subcutaneous injections. In general, however, they are thought to be safe and effective.

Finally, patients with moderate or severely active IBD may be treated with calcineurin inhibitors such as Cyclosporin A. Interest in Cyclosporine A stemmed from its initial use as an inhibitor of organ transplant rejection.[74] Its mechanism of action was found to involve inhibition of a transcription factor called NFAT, nuclear factor of activated T cells. NFAT is now a well-recognized regulator of T cell activation and differentiation as well as many functional aspects of dendritic cells, B cells, and metakaryocytes. Target genes of NFAT include a number of chemokines and cytokines such as FOXP3, IL2,IL4, IL10, IL13, IL17, and TNFα. Due to its potential for inducing kidney toxicity, its use is now less favored as compared to that of infliximub.

In considering whether either increased consumption of flavonoids or use of flavonoids as supplements would be useful to patients with a chronic disease state such as IBD, a number of factors are important. First and perhaps most importantly, would high consumption of flavonoids prevent the onset of IBD? Second, would the high consumption and/or supplemental administration of flavonoids enhance the effectiveness of drugs currently used to treat IBD? Third, could flavonoids be used to ameliorate the side effects associated with currently used IBD therapeutics? Fourth, could flavonoids be used to inhibit or slow the progression of IBD to colorectal cancer? Finally, by probing for mechanistic insights into how flavonoids exert their anti-inflammatory effects, could novel and more effective IBD therapeutics be developed? We will begin to address at least some of these questions as we explore the experimental tools and recent advances used to understand how flavonoids may impact IBD.

Animal models used to study IBD. Currently, over 50 different animal models of inflammatory bowel diseases are used.[75–78] These murine models are typically classified as chemically induced, genetically engineered (containing either transgenes or representing gene knock outs) and adoptive cell transfer models. Unfortunately, none of the models are considered adequate as pre-clinical models as they do not satisfy the requirements of a highly translatable system. These requirements include sharing similar pathological hallmarks of disease, underlying mechanisms and biology, and response to standard of care therapeutics with that of the human patient. The best prediction of human trial outcomes are obtained when a variety of pre-clinical models have been used to test a given therapeutic and yield similar results. Other difficulties encountered when developing novel treatments for IBD pertain to the lack of appropriate disease biomarkers, which limits our ability to translate the results obtained in the pre-clinical models to events observed in human subjects. It is also very difficult to identify the most likely responsive patients. The high genetic diversity of IBD, involving more than 99 non-overlapping genetic risk loci, further contributes to the difficulties encountered when anticipating clinical responses to a given agent. Finally, clinical trials are often performed in patients that are resistant to conventional therapies and thus may be similarly resistant to many types of therapeutic approaches.

In determining the best model to be used for testing whether a flavonoid may be effective in preventing IBD, the first consideration should focus on its predicted mechanism of action. Regardless of the model to be used, the measures of disease severity are similar. This includes loss of body weight and disease activity scores which incorporate the extent of diarrhea and presence of occult blood. Colon wet weight is often included in the macroscopic analyses of colitis as the weight of the inflamed tissue increases with disease severity. More specific measures of inflammation include the analyses of myeloperoxidase (MPO) activity, an indicator of infiltration of neutrophil granulocytes, expression levels of pro-inflammatory cytokines such as TNFα, IL1β, and IL6, and indicators of NFκB activity.

Chemically induced models include the use of dextran sodium sulfate (DSS) or trinitrobenzene sulfonic acid (TNBS). The administration of DSS is toxic to the epithelial layer of the intestinal mucosa and thus disrupts the intestinal barrier. This allows for the entry of commensal bacteria into the lamina propria which initiates the inflammatory response to the gut microorganisms. The acute DSS model is used quite frequently as it is cheap and easy to use, and the severity of disease can be modified by varying the concentrations of DSS administered in the drinking water. Typically, concentrations of 1.5% to 4% DSS are used. However, the chronic model of DSS may also be used. A typical chronic DSS study includes the administration of DSS for a seven-day period followed by a 14-day period (without DSS) to allow for rest and recovery. This cycle can be repeated several times. This is thought to mimic the cycle of acute inflammation and mucosal healing experienced by the IBD patient. It is commonly thought that in the acute DSS model, the most common immune cells present are neutrophils and macrophages and that the T cell response becomes more evident in the chronic model.

TNBS is a haptenizing agent that triggers an immune response. TNBS-induced colitis is characterized by diffuse colonic inflammation and increased leukocyte infiltration, edema, and ulceration. It is thought to be mediated primarily via a T_H1 immune response as indicated by increases in the expression levels of IL12 and TNFα, and mimics events that occur in patients with CD. However, in some strains of mice, evidence of a T_H2-mediated response can also be observed. TNBS is typically dissolved in alcohol and delivered rectally. Alcohol plays a role in initiating the immune response as it breaks down the mucosal barrier. Mouse strains responsive to TNBS-induced colitis include SJL/J, BALBC, and C57BL/6. A typical experimental paradigm involves rectal administration of 200 mg/kg of TNBS dissolved in 30% ethanol. The infiltration of immune cells is typically observed within two hours; symptoms of chronic inflammation occur within 48 hours; and the experiments are typically terminated after 72 hours.

A comparison of the acute DSS, chronic DSS, and TNBS models has reported that increases in IFNγ+, IL17+, but not IL4+ expressing cells (corresponding to T_H1, T_H17, and T_H2-mediated responses, respectively) can be detected in all three models.[79] A disadvantage noted using acute models of chemically induced colitis as compared to the genetic models is that a higher degree of variability is often observed.[76] This is proposed to be due to the shorter disease course.

Other chemically induced models of experimental colitis include the use of oxazolone, acetic acid, peptidoglycan, carrageenan, and non-steroidal anti-inflammatory agents.[78] Oxazolone is a haptenating agent that is thought to induce primarily a T_H2 response and is a proposed model of UC. Like TNBS, oxazolone is administered rectally with ethanol serving as the vehicle carrier. Typical experimental paradigms include first pre-sensitizing the mice by applying a 3% oxazolone solution to their shaved skin. After 5–8 days, a second dose of oxazolone (0.75–1%) is administered rectally. Symptoms of colitis develop and progress within 7–12 days. Acetic acid is also used to induce a form of colitis that is thought to closely resemble human IBD. It is administered rectally and is thought to induce colitis via induction of oxidative stress. In mice, a 4–5% solution of acetic acid in 0.9% saline is used. Symptoms of colitis are typically evident within four days of acetic acid administration. Colitis induced by peptidoglycan, carrageenan, or non-steroidal anti-inflammatory agents such as indomethacin is not as well characterized as those discussed above and will thus not be considered further.

With respect to genetically engineered models of IBD, two types are used: the transgenic and the knock-out mouse models.[75–77] In the transgenic model, the spontaneous colitis that arises is due to selective overexpression of specific genes that often intensify the effector cell response.

Examples include overexpression of IL17 in mucosal T cells or overexpression of the STAT4 regulator in CD4+ T cells. In the knock-out models, key genes involved in the regulatory response are often selectively targeted. Included here are models of defective mucosal barrier function, innate immunity, and adaptive immunity. Examples of genes often targeted include IL10, IL2, TGFβ, STAT3, and TNFAIP3 (a potent cytoplasmic inhibitor of NFκB).

The third type of model that is often used to study colitis in a laboratory setting is the adoptive transfer model. Here, either CD 4+ or CD 8+ T cells are transferred to an immune compromised mouse such as RAG$^{-/-}$ or SCID.[80] Exposure to commensal bacteria in the gut is thought to initiate the differentiation of the administered naïve T cells into primarily T_H1 and T_H17 cells. Chronic inflammation is typically observed after three to six days following transfer. This approach allows for the determination of the specific role of the innate versus adaptive immune cells or for specific T cell populations. For example, using this model, it has been demonstrated that the T_H1 and T_H17 cells enhance whereas T_{reg} cells suppress intestinal inflammation. The increasing use of the adoptive transfer model is thought to be due largely from observations that spontaneous colitis formed in the genetically engineered models lacks adequate characterization and often results in low disease penetrance. In comparison, the adoptive cell transfer model is purported to be more predictable, results in a high disease penetrance, and occurs in a strain-independent manner.

A number of additional factors should also be considered when designing experiments for testing the potential effectiveness of a given flavonoid on preventing gut inflammation.[75,76] Gender-specific effects have been reported when using both chemically induced and genetic models of IBD. Female mice may be preferred over males as males often engage in aggressive behavior that will increase the stress and inflammatory responses. On the other hand, the higher levels of 17β estradiol in the females may exert either pro-inflammatory or anti-inflammatory effects, depending on the model that is being used.[81] Strain differences also exist.[76,82] For example, the SJL and BALB/c mice are more susceptible and the C57BL6 mice more resistant to disease induced by TNBS. With respect to DSS-induced colitis, however, the most sensitive strains include C3H/HeJBir and C57BL/6J, while the least sensitive strain is DBA/2J.

The timing of flavonoid administration with respect to disease onset should also be considered. For example, pre-treatment with the flavonoid would potentially impact events associated with the development of inflammation whereas post-treatment would alter events associated with its ability to attenuate pre-existing inflammation. Other factors to be considered include the diet, as commercially available preparations often contain flavonoids that are constituents of alfalfa. Finally, sources of variation are attributed to the commercial source of the mice and their commensal intestinal microbiome.

An example of experimental data that is often generated using the DSS-induced model of experimental colitis is shown in Figure 3.1. Here, two different concentrations of DSS were administered to the drinking water of male, C57BL/6J mice. Unadulterated tap water served as the negative control (water only). A primary symptom of colitis, weight loss, typically becomes evident at 4–5 days following the administration of DSS. Additional symptoms, such as loose stool, also begin to appear which become more severe as time progresses and can be used to tabulate a disease severity index. DSS-induced inflammation in the colon is apparent by the increase in colon weight and colon shortening. As can be seen in Fig. 3.1, colitis induced by 3% DSS is more severe than that induced by 1.5% DSS as indicated by the greater loss of weight and shorter colon weight. Changes in the expression patterns of pro-inflammatory cytokines typically observed in the DSS model are shown in the lower panel of Fig. 3.1. The most highly induced cytokines in the DSS model are typically IL6, KC (the murine IL8 homolog), and IL17A.

The value of using multiple mouse models may be best illustrated in the body of work that has developed over the years to understand the role of the NFκB pathway in intestinal homeostasis and the pathogenesis of IBD.[83,84] Excessive NFκB signaling contributes to the pathogenesis of IBD and thus, NFκB inhibitors such as curcumin are thought to be therapeutically beneficial. However, a study of mice harboring a variety of genetic modifications that impinge on the NFκB pathway has revealed that the interplay that exists between bacterial sensing and the inflammatory responses mediated by the innate and adaptive immune responses regulated by NFκB is intricate and often difficult to predict. It is a system of constant surveillance that is honed by the absence or presence of particular intestinal microorganisms, adapted by the host organism often via

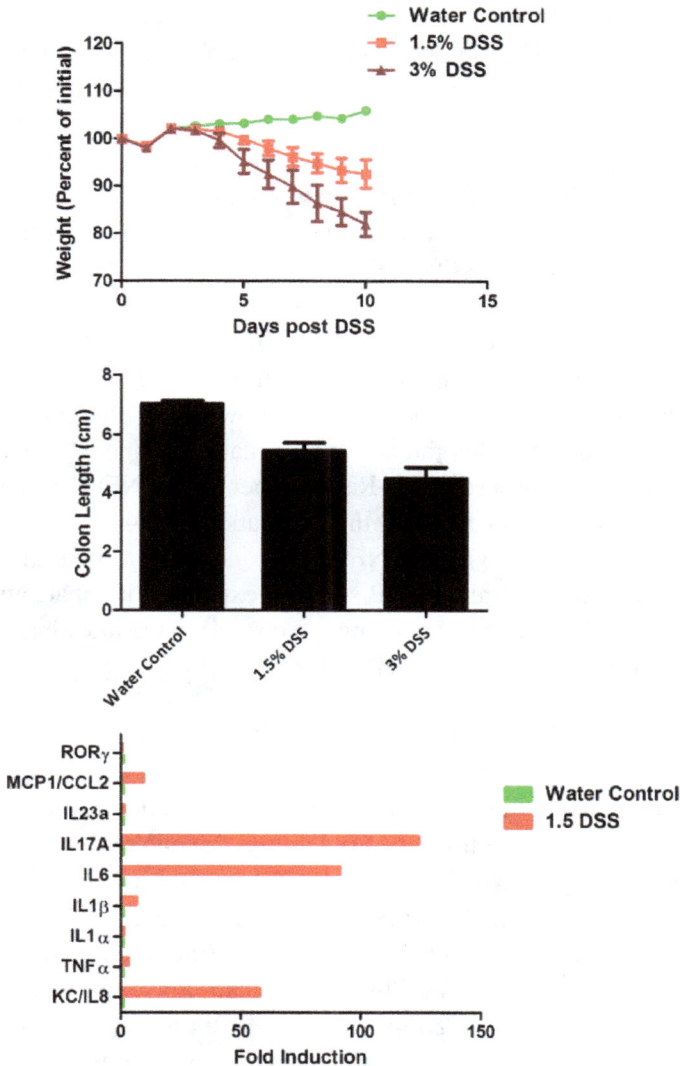

Fig. 3.1. A representative DSS-induced colitis experiment. Either 1.5% or 3% DSS was administered to the drinking water of male C57BL/6J mice. A third group was administered water (water control). After seven days, all groups were administered water for an additional threedays.The mice were then euthanized and the tissues were collected and analyzed. The mRNA expression of the indicated genes from the pooled samples was measured in the colon tissue using real time quantitative polymerase chain reactions and normalized to those of water control. The top graph shows the effect of DSS on mouse weights. The middle graph shows the effect of DSS on colon weights. The bottom graph shows the effect of 1.5% DSS on expression of cytokines/transcription factors. The results are representative of 5–11 mice (*Unpublished data from the Swanson Laboratory*).

opposing but interlocking responses by the innate and adaptive immune cells. For example, TLR signaling in the macrophages, when appropriate, can produce a signal that enables epithelial cells to proliferate after DSS damage and thereby contribute to mucosal healing. However, this TLR signaling that arises following exposure to the intestinal bacteria excessively activates NFκB and is a major determinant of the colitis that develops in the IL10$^{-/-}$ mice. The key source of excessive NFκB in this model is the myeloid cells. Excessive NFκB signaling in the intestinal epithelial cells results in an increased susceptibility to damage-induced colitis (i.e., DSS-induced). However, complete absence of NFκB in these cells can also be deleterious due to impaired production of anti-microbial peptides that weaken the mucosal barrier thereby allowing for the entry of commensal bacteria and a heightened inflammatory response. Within the adaptive immune response, the c-Rel member of the NFκB family plays an important role in promoting the T_H1 and T_H17-mediated immune response. It is now thought that NFκB does not play a broad, protective role, but rather that different NFκB subunits expressed in innate and adaptive immune cells may exert opposing effects during the intestinal immune response and inflammatory bowel disease.

Impact of flavonoids on experimental colitis. Flavonoids that have been shown to impact experimental colitis using a variety of *in vivo* approaches include members of the flavones, flavanone, flavonol, isoflavone, and catechin subclasses. The impact of individual flavonoids in different animal models of experimental colitis will be discussed in detail as follows.

With respect to the flavones, the best studied are constituents of *Scutellaria baicalin*, a flowering plant used in traditional Chinese medicine.[85] It is also referred to as Chinese skullcap. Flavonoids identified in extracts of *Scutellaria baicalin* include apigenin, baicalin, baicalein, wogonin, and luteolin. Keeping in mind the observation that a major limitation associated with the *in vivo* administration of apigenin is its low water solubility, Apigenin K, a form of apigenin with enhanced water solubility, was generated.[86] It was then administered to female rats via a catheter and its efficacy with respect to inhibiting TNBS-induced colitis was evaluated. The impact of Apigenin K was compared to that of budesonide (2 mg/kg), a commonly used drug to treat inflammatory conditions of IBD. Apigenin K was tested at three doses (1 mg/kg, 3 mg/kg, and 10 mg/kg). The 3 mg/kg dose was found to be the most effective and resulted in reductions in measures of colon inflammation and expression levels of IL1β, TNFα, and IL6 that were similar to those observed using the positive control budesonide. While Apigenin K also appeared to be effective at inhibiting DSS-induced colitis, statistically significant differences were only observed with respect to IL1β expression.

The impact of baicalin has been examined using DSS-induced colitis in female mice relative to treatment with mesalazine.[85] Baicalin and mesalazine were administered twice daily via intragastric injections using a relatively high concentration,100 mg/kg. The treatments were initiated after colitis had been induced and continued for seven days. While neither baicalin nor mesalazine altered the induction of colitis, both significantly aided disease recovery to similar extents. The mechanisms underlying baicalin's effects involve suppression of TLR4; increased expression of IL10; decreased expression of TNFα, IL6, and IL13; and decreased expression of the NFκB subunit, p64, in the colon tissue. Similar results were obtained using TNBS-induced colitis in male rats.[87] Here, the doses of baicalin used were 20 mg, 10 mg, and 5 mg administered via gastric lavage, and were compared to that of mesalazine (100 mg/kg). Again, baicalin inhibited all markers of colitis (including measures of NFκB activation and expression levels of IL1β, TNFα, and IL6) in a dose-responsive manner with the highest dose nearly comparable to that of mesalazine. Baicalin has also been shown to be effective in inhibiting the proliferation and expression levels of pro-inflammatory cytokines (IL6 and IFNγ) in cultured peripheral blood mononuclear cells isolated from patients with UC.

More recent studies have contributed to our understanding of how baicalin may exert its anti-inflammatory effects.[88] Here, the actions of baicalein versus baicalin, its glucuronide metabolite that is likely formed in the intestine, were compared. Baicalein, but not baicalin, was found to be a relatively potent activator of the steroid hormone receptor PXR. The induction of PXR by baicalein appears to be mediated via its induction of Cdx2 (an intestine-specific homeobox protein, caudal type homeobox 2). Cdx2 is a central regulator of intestinal cell differentiation and intestine homeostasis, and regulates the expression levels of PXR. Baicalein, but not baicalin, upregulates Cdx2 protein levels that in turn, upregulates the expression of PXR. Baicalein, and to a lesser extent, baicalin, when administered orally, alleviated DSS-induced colitis in mice in a PXR-dependent manner. Thus, it appears that the anti-inflammatory effects of baicalin require its conversion to the aglycone baicalein and inhibit the NFκB pathway via enhancing its negative regulator PXR.

The impact of luteolin on experimental colitis is somewhat contradictory, but appears to be dependent on the model used to study its effects.[89,90] Nishitani *et al.* administered three doses of luteolin (5 mg/kg, 20 mg/kg and 50 mg/kg) via intragastric lavage for seven days prior to DSS administration and for the duration of the experiment.[89] Female mice were used. At the higher doses used (20 mg/kg and 50 mg/kg) luteolin was found to inhibit all macroscopic measures of colitis (i.e., colon length, colon histology, etc.). Analyses of pro-inflammatory cytokines, however, revealed that only the expression levels of IFNγ were significantly decreased. Karrasch *et al.*, however, reported that when 2% luteolin was administered in the chow (daily consumption was estimated to be 32 mg of luteolin), the luteolin-treated mice lost more weight and exhibited greater disease severity after the administration of DSS.[90] In this study, the mice expressed the green fluorescent reporter gene that was controlled by NFκB. Thus, they were able to directly observe that luteolin inhibited NFκB activity in the enterocytes. Using the same dietary regimen, the impact of luteolin was observed for four weeks in IL10$^{-/-}$ mice. These mice are thought to model a chronic form of T_H1-mediated inflammation. Here, luteolin alleviated all indicators of colitis. The authors of this study postulated that luteolin inhibits the protective function of NFκB, thereby enhancing the intestinal cell injury that is induced by DSS and exacerbating the ensuing inflammatory response. In the IL10$^{-/-}$ mouse, however, luteolin's inhibition of NFκB resides primarily within the lamina propria where it inhibits pro-inflammatory cytokine secretion. It remains to be determined, however, whether the route by which luteolin is administered may also play a role. That is, when administered via the diet, the exposure to luteolin would be somewhat constant as compared to the more intermittent exposure provided by gastric administration.

The best characterized flavanones tested in experimental models of colitis are liquiritigenin and naringenin.[91–94] The impact of liquiritigenin has been studied using TNBS-induced colitis. Here, two doses of liquiritigenin (10 mg/kg and 20 mg/day) were administered daily by oral gavage for three days, colitis was induced, and the experiments were terminated after 24 hours. Mesalazine, administered at a dose of 10 mg/kg, served as the positive control. Liquiritigenin treatment alleviated the majority of TNBS-induced changes observed, including macroscopic measures, expression levels of pro-inflammatory cytokines, and phosphorylation of p65 (as a measure of NFκB activity). The most consistent improvements were observed at the 20 mg/kg dose of liquiritigenin which suppressed inflammation at an extent that was similar to that of mesalazine.

The impact of naringenin on experimental colitis has been shown in three separate studies.[92–94] Al-Rejaie *et al.* tested the impact of naringenin on acetic acid-induced colitis in male rats.[92] Naringenin was administered in three different daily doses (25 mg/kg, 50 mg/kg, and 100 mg/kg) by gavage for seven days, colitis was induced, and the experiments were terminated after 24 hours. Mesalazine (300 mg/kg) served as the positive control. Naringenin pre-treatment alleviated acetic acid-induced macroscopic measures of colitis in a dose-dependent manner. At the highest dose (100 mg/kg), the effect of naringenin pre-treatment was similar to that observed in the mesalazine-treated rats. The expression levels of pro-inflammatory cytokines as well as measures of oxidative stress induced by acetic acid were similarly attenuated by naringenin treatment. A second study was performed using DSS-induced colitis in male mice.[93] Here, naringenin (0.3%) was administered in the food for the duration of the DSS treatment. Naringenin protected against all macroscopic measures of DSS-induced colitis and DSS-induction of pro-inflammatory cytokines (i.e., IFNγ, IL6, and IL17A). Further analyses of intestinal permeability indicated that naringenin may exert protective effects toward the maintenance of intestinal barrier function. Finally, the effects of naringenin on DSS-induced colitis in female mice were evaluated.[94] Here, naringenin was administered by daily gavage (50 mg/day) for three days prior to, and for the duration of, the study. Naringenin treatment alleviated the macroscopic changes elicited by DSS administration as well as DSS-induction of TLR4 expression, activation of NFκB (as indicated by nuclear localization and phosphorylation of p65), and expression levels of pro-inflammatory cytokines.

Amongst the flavonols, the best characterized with respect to its impact on experimental colitis is quercetin and its glycosides rutin and quercitrin.[95–98] A series of studies has been performed to ascertain the effectiveness and underlying mechanisms of quercetin with respect to its ability to inhibit colonic inflammation. Using daily doses of quercitrin that ranged from 0.125 mg/kg to 25 mg/kg, Sanchez de Medina *et al.*, found that two doses, 1 mg/kg and 5 mg/kg, were effective in inhibiting TNBS-induced colitis in female rats.[95,99] In this acute study, quercitrin was administered two hours prior to the induction of colitis and after 24 hours. Measures of colitis were evaluated 48 hours after induction. In a chronic study of TNBS colitis, the rats were administered quercitrin (1 or 6 mg/kg/day) and the tissues were analyzed 2–4 weeks after colitis was induced. At the lower dose (1 mg/kg/day), quercitrin appeared to alleviate at least some measures of chronic colitis, such as colonic absorption. In a follow-up study, examination of the effects of quercitrin at the early stages (i.e., 24 hours after TNBS administration) indicated that quercitrin counteracted TNBS-induced oxidative stress, but not inflammation.[96] The anti-inflammatory effects of quercitrin were then assessed using the DSS-induced model of colitis.[99] Here, female rats were orally administered quercitrin (1 mg/kg/day and 5 mg/kg/day) either prior to or following the administration of DSS to evaluate its protective and therapeutic potential, respectively. At the 1 mg/kg/day dose, quercitrin attenuated measures of DSS-induced colitis when administered prior to DSS exposure. This included attenuation of DSS-induced expression of pro-inflammatory cytokines, NFκB activity, and iNOS activity in the colon. Quercitrin also provided a beneficial and protective effect when administered after colitis was established. However, when evaluated for anti-inflammatory properties *in vitro*, quercetin, but not its glycoside quercitrin, was found to be highly potent.[97] Using the DSS-induced model of colitis in female rats, the *in vivo* impact of quercitrin versus quercetin was then compared. Quercitrin or quercetin was administered at an oral dose of 1 mg/kg/day for the last ten days of DSS administration. Quercitrin, but not quercetin, attenuated DSS-induced colitis in a manner that was apparent seven days after DSS treatment was started. Measures of colitis that were attenuated by quercitrin included colonic expression levels of pro-inflammatory cytokines, TNFα, and IL1β, as well as iNOS. The authors also demonstrated that quercetin could be released from the aglyconic form and thereby exert its anti-inflammatory effects. What these results imply is that the oral administration of quercetin is unlikely to be effective due to its low bioavailability. Oral administration of the glycoside quercitrin, however, is effective because it survives the acidity of the gut and is released to its pharmacologically active form quercetin by the fecal bacteria.

Other studies have similarly examined the anti-inflammatory properties of either derivatives of quercetin or encapsulated forms of quercetin.[100–102] Using a rat model of acetic acid-induced colitis, the anti-inflammatory effects of quercetin, chloronaphthoquinone quercetin, or monochloropivaloyl quercetin were examined.[100] Of these, only chloronapthoquinone quercetin was found to be effective in inhibiting markers of acetic acid colitis. The *in vivo*, anti-inflammatory effects of quercetin have also been improved using microcapulation techniques. Guazelli *et al.* loaded quercetin onto pectin/casein microcapsules which allowed for the slow and graduate release of quercetin in the gut.[101] Using the acetic acid model of colitis in male mice, a dose-dependent (1 mg/kg, 10 mg/kg, and 100 mg/kg) improvement in all measures of acetic acid-induced colitis was observed. This included quercetin alleviation of pro-inflammatory cytokine expression (IL1β and IL33) and protection against acetic acid-reduced antioxidant levels in the colon. Similar results have been obtained using quercetin encapsulated within polyethylene glycol.[102]

Studies that have examined the effects of rutin (3-O-rhamnosylglycosyl quercetin) in models of experimental colitis have similarly shown that this glycoside is more effective than the parent quercetin.[98] Kwon *et al.* used the DSS-induced model of colitis and female mice to compare the effects of oral administration of quercetin versus that of rutin. The doses used ranged from 60 μg/day to 6 mg/day and were administered either prior to or following the administration of DSS. Rutin, but not quercetin, at the highest dose tested attenuated DSS-induced morphological changes and colonic expression of pro-inflammatory cytokines (i.e., IL1β and IL6). When administered therapeutically (i.e., after DSS administration), treatment with rutin, but not quercetin, similarly reduced markers of DSS-induced colitis. In rats, rutin has been shown to be deglycosylated in the fecal contents to liberate quercetin.[103] Oral administration of rutin (10 mg/kg) was found to result in an accumulation of quercetin (~200 μM) in the large intestine. In the TNBS-induced rat model of colitis, oral administration of rutin 10 mg/kg/day was comparable to that of sulfasalazine (30 mg/kg) in preventing TNBS-induction of colon damage and myeloperoxidase activity. Finally, rutin has also been shown to effectively inhibit colitis induced by the cell transfer method.[104] Here, naïve T cells (CD4+CD62L+) isolated from female C57BL/6 mice were transferred into C57BL/6 Rag1[-/-] mice. Treatment with rutin was initiated after eight weeks when disease symptoms became evident. Rutin was administered by gavage at either 28.5 mg/kg/day or 57 mg/kg/day. This latter dose corresponds to approximately 448 mg of rutin in humans. Oral administration of budesonide (3 mg/kg/day) served as a positive control. The lowest dose of rutin (28.5 mg/kg/day) did not impact any measure of colitis after 22 days of observation. At 57 mg/kg/day and after 12 days, oral administration of rutin attenuated all measures of colitis including colonic expression levels of pro-inflammatory cytokines (IFNγ, TNFα, CXCL1, IL1β, IL6, and IL17) as well as COX-2. The levels of activated STAT4 in the colon were also reduced following treatment with rutin. The protective effects presented by rutin were comparable to that of budesonide. The authors of this study propose that the effects of rutin involve release of quercetin which then acts primarily on the IFNγ-producing cells such as the T_H1 lymphocytes. Taken together, these studies indicate that quercetin maybe an effective anti-inflammatory agent if its bioavailability is improved using approaches such as microencapsulation delivery or by administering its glycosidic form.

With respect to isoflavones, the best characterized in experimental models of colitis are equol, genistein, and daidzein.[105–108] To determine the impact of prolonged exposures to isoflavone phytoestrogens, females rats were fed diets that were either isoflavone-enriched or isoflavone-depleted while pregnant and during lactation.[106] The isoflavone-enriched diets contained 240 µg/g genistein and 232 µg/g daidzein. Upon weaning, the male pups were maintained on these respective diets and when 11 weeks of age, were subjected to TNBS colitis. The isoflavone-enriched diets appeared to increase the severity of colitis as indicated by an increase in TNBS-induced colon wet weight and myeloperoxidase activity. However, the levels of COX-2 expression was reduced in the TNBS-treated rats maintained on the isoflavone-rich diet as compared to that observed in the isoflavone-poor group. A second study used a "therapeutic paradigm" to test the effect of genistein on TNBS-induced colitis in male rats.[108] Here, genistein (100 mg/kg) was administered daily via gavage after TNBS colitis was initiated. This dosing regimen resulted in the expected increase in weights of prostate and seminal vesicles due to the previously reported estrogenic effects of genistein. With respect to measures of TNBS-induced colitis, genistein treatment did not impact colon wet weight. However, it did significantly decrease markers of inflammation in the colon tissue, in particular, that of myeloperoxidase activity and COX-2 expression. Finally, a study to examine the impact of soy isoflavones on DSS-induced colitis in female mice was reported.[105] Here, either genistein, daidzein, or equol was administered at daily doses of 20 mg/kg for seven days prior to and five days following the administration of DSS. Analyses of DSS-induced decreases in body weight revealed that treatment with genistein had no effect, but daidzein and equol treatments exacerbated the effects of DSS. A dose response experiment indicated that the dose of equol required to elicit this detrimental effect was 20 mg/kg. Analyses of cytokines produced in the mesenteric lymph nodes indicated that only IL10 levels were significantly affected by equol and were decreased relative to the DSS control. Interestingly, a recent epidemiology study of Japanese patients indicated that in females, a high consumption of isoflavones is associated with increased risk of developing UC.

While several catechins have been examined with respect to their ability to modulate symptoms of experimental colitis, EGCG appears to be the best characterized.[109–115] A study examining the impact of a diet high in catechins (0.05%) and alpha tocopherol (0.25%) on TNBS-induced colitis in male rats was reported.[109] The diet was administered for one week after colitis induction. As compared to the control diet, the supplemented diet protected against TNBS-induced decreases in body weight, TNBS-induced increases in colon wet weight, and TNBS-induced increases in myeloperoxidase activity. The impact of the arubigin in TNBS-induced colitis in female mice was also examined.[111] Increasing doses of the arubigin (10–100 mg/kg/day) were administered orally ten days prior to the induction of colitis and eight days there after. Improved measures of TNBS-induced colitis (i.e., body and colon weights) were observed at the higher doses of the arubigin (40 and 100 mg/kg). In addition, these doses of the arubigin protected against TNBS-induced nitric oxide and superoxide production in the neutrophils as well as decreased TNBS-induced increases in myeloperoxidase activity and iNOS levels in the colon tissue. Markers of pro-inflammatory cytokines (i.e., IFNγ and IL12) and reduced activation of NFκB were also observed in the colon tissues from the the arubigin+ TNBS-treated mice as compared to those administered only TNBS. The TNBS-induced model of colitis performed in female mice has also been used to study the impact of theaflavin-3,3'digallate.[112] Theaflavin-3,3' digallate was administered orally at concentrations ranging from 1 to 10 mg/kg/day. The dosage regimen was initiated ten days prior to the administration of TNBS and for eight days thereafter. Theaflavin-3,3' digallate exerted a dose-response decrease in all measures of TNBS-induced colitis but reached significance at the higher (5 and 10 mg/kg) doses. Colon levels of myeloperoxidase activity, TNFα, IFNγ, IL12, iNOS, and activated NFκB were decreased in the groups treated with Theaflavin-3,3' digallate+TNBS versus those treated only with TNBS.

A number of studies have determined the effects of EGCG using a variety of models of experimental colitis.[110,113–117] EGCG (10 mg/kg administered twice per day via intraperitoneal injection) was administered to mice following their exposure to TNBS and their tissues were examined after one, three, and seven days.[110] Treatment with EGCG diminished the severity of TNBS-induced colitis at all time points examined, but was most prominent at the earlier time points. While TNBS-induced myeloperoxidase levels were reduced by EGCG, plasma levels of TNFα, IL6, IL10, and KC were not affected, indicating that EGCG's effects were not systemic. The authors speculated that EGCG prevents early oxidative stress that occurs during the initiation of colitis. Further examination of EGCG's impact on the early events of colitis (using the TNBS-induced model in rats) revealed that EGCG reduced histamine levels in the colon, an indication that it inhibits mast cell activity.[113] The effects of EGCG were also examined using acetic acid-induced colitis in rats.[115] EGCG was administered via daily gavage, after their exposure to acetic acid at a dose of 50 mg/kg/day and for seven days thereafter. The effects of EGCG were compared to those of sulfasalazine (0.25 g/kg/day). All macroscopic measures of colitis indicated that EGCG was more effective than sulfasalazine at attenuating colitis. EGCG also inhibited acetic acid-induced serum cytokine expression (TNFα and IFNγ), indicators of oxidative stress (i.e., malondialdehyde), and upregulated indicators of antioxidant activation (i.e., superoxide dismutase activity).

Efforts to improve the effectiveness of EGCG by enhancing its bio-availability include the development of an acetylated form (peracety-lated EGCG).[114] The effect of peracetylated EGCG versus EGCG was tested in DSS-induced colitis in male mice when incorporated into the diet (0.01% and 0.05% EGCG versus 0.17% and 0.85% peracetylated EGCG). Peracetylation of EGCG resulted in a six-fold higher concentration of EGCG in the colon. While both forms of EGCG were found to attenuate all measures of DSS-induced colitis, the peracetylated form was more effective than EGCG. Peracetylated EGCG reduced DSS-induced colon levels of iNOS, COX-2, IL6, and TNFα, but not IL1β. Further, peracetylated EGCG inhibited DSS-induced activation of NFκB while upregulating NRF2 and the expression levels of the NRF2 target gene and antioxidant, heme oxygenase. Additional attempts to improve the bioavailability of catechins include the use of oligonol, a standardized formulation consisting of 17.6% catechin-type monomers and 18.6% dimers and trimers in DSS-induced colitis.[118] Here, the oligonol formulation proved to be effective in inhibiting all measures of DSS-induced colitis including activation of NFκB and STAT3. In addition, a number of antioxidant genes were upregulated in the oligonol/ DSS treated mice.

A single study indicates, however, that high levels of EGCG can have detrimental effects.[117] Here, EGCG was incorporated into the diet at concentrations of 0.1%, 0.3%, and 0.5% EGCG. The female mice were maintained on EGCG-containing diets for one week prior to the administration of DSS and for the duration of the experiment. At the highest concentration tested (0.5%), EGCG treatment enhanced bleeding, enhanced DSS-induced body weight loss, and did not prevent DSS-induced colon shortening. Evaluation of inflammatory markers (i.e., serum levels of prostaglandin E2 and leukotriene B4) indicated that the high dose of EGCG exacerbated DSS-induced systemic inflammation. The doses of EGCG (0.3 and 0.5%) represent approximately 1.5g and 2.5g of EGCG in a 2,000 kcal diet consumed by a human subject.

A randomized, double blinded placebo-controlled phase 2a pilot study indicates that EGCG supplementation may be of value for patients with IBD.[116] Here, a total of 20 patients were randomly assigned to receive either a placebo control, or Polyphenol E® capsules containing either 200 mg or 400 mg EGCG twice daily. All patients had been diagnosed with UC and were taking 5-aminosalicylic when they enrolled in the study. The effectiveness of Polyphenol E® is indicated by the 66.7% response rate (10 out of 15 patients) as compared to 0% (0 out of 4 patients). The active remission rate was 53.3% versus 0% for the Polyphenol E® and placebo control groups, respectively.

Taken together, with the results obtained using different animal models of experimental colitis combined with an early clinical trial, the use of EGCG as a therapeutic for treating IBD is promising. Similar approaches may prove to be useful for the future development of additional flavonoid-based therapies.

References

1. Peakman M. *Basic and Clinical Immunology*: Elsevier Health Sciences; 2009.
2. Bamias G, Arseneau KO, Cominelli F. Cytokines and mucosal immunity. *Curr Opin Gastroenterol.* Nov 2014;30(6):547–552.
3. Bulek K, Swaidani S, Aronica M, Li X. Epithelium: The interplay between innate and Th2 immunity. *Immunol Cell Biol.* Mar–Apr 2010;88(3):257–268.
4. Jovanovic K, Siebeck M, Gropp R. The route to pathologies in chronic inflammatory diseases characterized by T helper type 2 immune cells. *Clin Exp Immunol.* Nov 2014;178(2):201–211.
5. De Simone V, Pallone F, Monteleone G, Stolfi C. Role of T17 cytokines in the control of colorectal cancer. *Oncoimmunology.* Dec 2013;2(12):e26617.
6. Murray PJ, Wynn TA. Protective and pathogenic functions of macrophage subsets. *Nature Reviews:Immunology.* Nov 2011;11(11):723–737.
7. Mariani F, Sena P, Roncucci L. Inflammatory pathways in the early steps of colorectal cancer development. *World Journal of Gastroenterology.* Aug 2014;20(29):9716–9731.
8. Schaefer L. Complexity of danger: The diverse nature of damage-associated molecular patterns. *J BiolChem.* Dec 2014;289(51):35237–35245.

9. Pal S, Bhattacharjee A, Ali A, Mandal NC, Mandal SC, Pal M. Chronic inflammation and cancer: Potential chemoprevention through nuclear factor kappa B and p53 mutual antagonism. *J Inflamm.* 2014;11:23.

10. Elinav E, Nowarski R, Thaiss CA, Hu B, Jin C, Flavell RA. Inflammation-induced cancer: Crosstalk between tumours, immune cells and microorganisms. *Nature Reviews:Cancer.* Nov 2013;13(11):759–771.

11. McAllister SS, Weinberg RA. The tumour-induced systemic environment as a critical regulator of cancer progression and metastasis. *Nature Cell Biol.* Aug 2014;16(8):717–727.

12. Roxburgh CS, McMillan DC. Cancer and systemic inflammation: Treat the tumour and treat the host. *Br J Cancer.* Mar 2014;110(6):1409–1412.

13. Landskron G, De la Fuente M, Thuwajit P, Thuwajit C, Hermoso MA. Chronic inflammation and cytokines in the tumor microenvironment. *J Immunol Res.* 2014;2014:149185.

14. Shaulian E. AP-1 — The Jun proteins: Oncogenes or tumor suppressors in disguise? *Cellular Signalling.* Jun 2010;22(6):894–899.

15. Vaiopoulos AG, Papachroni KK, Papavassiliou AG. Colon carcinogenesis: Learning from NF-kappaB and AP-1. *Int J BiochemCell Biol.* Jul 2010;42(7):1061–1065.

16. Yu H, Lee H, Herrmann A, Buettner R, Jove R. Revisiting STAT3 signalling in cancer: New and unexpected biological functions. *Nature Reviews:Cancer.* Nov 2014;14(11):736–746.

17. Mamlouk S, Wielockx B. Hypoxia-inducible factors as key regulators of tumor inflammation. *Int J Cancer.* Jun 2013;132(12):2721–2729.

18. Leinonen HM, Kansanen E, Polonen P, Heinaniemi M, Levonen AL. Role of the Keap1-Nrf2 pathway in cancer. *Adv Cancer Res.* 2014;122: 281–320.

19. Wakabayashi N, Slocum SL, Skoko JJ, Shin S, Kensler TW. When NRF2 talks, who's listening? *Antioxid Redox Signal.* Dec 2010;13(11):1649–1663.

20. Murray IA, Patterson AD, Perdew GH. Aryl hydrocarbon receptor ligands in cancer: Friend and foe. *Nature Reviews:Cancer.* Nov 2014;14(12):801–814.

21. Hanieh H. Toward understanding the role of aryl hydrocarbon receptor in the immune system: Current progress and future trends. *BioMed Research International.* 2014;2014: Article ID 520763.

22. Banerjee M, Robbins D, Chen T. Targeting xenobiotic receptors PXR and CAR in human diseases. *Drug Discovery Today.* Nov 20, 2014.

23. Bionaz M, Hausman GJ, Loor JJ, Mandard S. Physiological and nutritional roles of PPAR across species. *PPAR Research.* 2013;2013:Article ID 807156.

24. Johnson JL, de Mejia EG. Flavonoid apigenin modified gene expression associated with inflammation and cancer and induced apoptosis in human pancreatic cancer cells through inhibition of GSK-3beta/NF-kappaB signaling cascade. *Mol Nutr Food Res.* Dec 2013;57(12):2112–2127.

25. Dou W, Zhang J, Zhang E, *et al.* Chrysin ameliorates chemically induced colitis in the mouse through modulation of a PXR/NF-kappaB signaling pathway. *J Pharmacol Exp Ther.* Jun 2013;345(3):473–482.

26. Zhao H, Zhang X, Chen X, *et al.* Isoliquiritigenin, a flavonoid from licorice, blocks M2 macrophage polarization in colitis-associated tumorigenesis through downregulating PGE2 and IL-6. *Toxicol Appl Pharmacol.* Sep 2014;279(3):311–321.

27. Yang JH, Kim SC, Shin BY, *et al.* O-Methylated flavonol isorhamnetin prevents acute inflammation through blocking of NF-kappaB activation. *Food Chem Toxicol.* Sep 2013;59:362–372.

28. Chen X, Yang X, Liu T, *et al.* Kaempferol regulates MAPKs and NF-kappaB signaling pathways to attenuate LPS-induced acute lung injury in mice. *Int Immunopharmacol.* Oct 2012;14(2):209–216.

29. Weng Z, Patel AB, Vasiadi M, Therianou A, Theoharides TC. Luteolin inhibits human keratinocyte activation and decreases NF-kappaB induction that is increased in psoriatic skin. *PLoS One.* 2014;9(2):e90739.

30. Chao CL, Weng CS, Chang NC, Lin JS, Kao ST, Ho FM. Naringenin more effectively inhibits inducible nitric oxide synthase and cyclooxygenase-2 expression in macrophages than in microglia. *Nutr Res.* Dec 2010;30(12):858–864.

31. Granado-Serrano AB, Martin MA, Bravo L, Goya L, Ramos S. Quercetin attenuates TNF-induced inflammation in hepatic cells by inhibiting the NF-kappaB pathway. *Nutr Cancer.* 2012;64(4):588–598.

32. Michaud-Levesque J, Bousquet-Gagnon N, Beliveau R. Quercetin abrogates IL-6/STAT3 signaling and inhibits glioblastoma cell line growth and migration. *Exp Cell Res.* May 2012;318(8):925–935.

33. Ruiz PA, Haller D. Functional diversity of flavonoids in the inhibition of the proinflammatory NF-kappaB, IRF, and Akt signaling pathways in murine intestinal epithelial cells. *J Nutr.* Mar 2006;136(3):664–671.

34. Ruiz PA, Braune A, Holzlwimmer G, Quintanilla-Fend L, Haller D. Quercetin inhibits TNF-induced NF-kappaB transcription factor recruitment to proinflammatory gene promoters in murine intestinal epithelial cells. *J Nutr.* May 2007;137(5):1208–1215.

35. Manna SK, Aggarwal RS, Sethi G, Aggarwal BB, Ramesh GT. Morin (3,5,7,2',4'-Pentahydroxyflavone) abolishes nuclear factor-kappaB activation induced by various carcinogens and inflammatory stimuli, leading to suppression of nuclear factor-kappaB-regulated gene expression and up-regulation of apoptosis. *Clin Cancer Res.* Apr 2007;13(7):2290–2297.

36. Yang JH, Shin BY, Han JY, *et al.* Isorhamnetin protects against oxidative stress by activating Nrf2 and inducing the expression of its target genes. *Toxicol Appl Pharmacol.* Jan 2014;274(2):293–301.

37. Zhai X, Lin M, Zhang F, *et al.* Dietary flavonoid genistein induces Nrf2 and phase II detoxification gene expression via ERKs and PKC pathways and protects against oxidative stress in Caco-2 cells. *Mol Nutr Food Res.* Feb 2013;57(2):249–259.

38. Sun GB, Sun X, Wang M, *et al.* Oxidative stress suppression by luteolin-induced heme oxygenase-1 expression. *Toxicol Appl Pharmacol.* Dec 2012;265(2):229–240.

39. Cuadrado A, Martin-Moldes Z, Ye J, Lastres-Becker I. Transcription factors NRF2 and NF-kappaB are coordinated effectors of the Rho family, GTP-binding protein RAC1 during inflammation. *J Biol Chem.* May 2014;289(22): 15244–15258.

40. Lim HA, Lee EK, Kim JM, *et al.* PPARgamma activation by baicalin suppresses NF-kappaB-mediated inflammation in aged rat kidney. *Biogerontology.* Apr 2012;13(2):133–145.

41. Feng X, Qin H, Shi Q, *et al.* Chrysin attenuates inflammation by regulating M1/M2 status via activating PPARgamma. *Biochem Pharmacol.* Jun 2014; 89(4):503–514.

42. Mahmoud AM. Hesperidin protects against cyclophosphamide-induced hepatotoxicity by upregulation of PPARgamma and abrogation of oxidative stress and inflammation. *Can J Physiol Pharmacol.* Sep 2014;92(9):717–724.

43. Yao J, Pan D, Zhao Y, *et al.* Wogonin prevents lipopolysaccharide-induced acute lung injury and inflammation in mice via peroxisome proliferator-activated receptor gamma-mediated attenuation of the nuclear factor-kappaB pathway. *Immunology.* Oct 2014;143(2):241–257.

44. Wang HK, Yeh CH, Iwamoto T, Satsu H, Shimizu M, Totsuka M. Dietary flavonoid naringenin induces regulatory T cells via an aryl hydrocarbon receptor mediated pathway. *J Agric Food Chem.* Mar 7 2012;60(9):2171–2178.

45. Gupta SC, Phromnoi K, Aggarwal BB. Morin inhibits STAT3 tyrosine 705 phosphorylation in tumor cells through activation of protein tyrosine phosphatase SHP1. *Biochem Pharmacol.* Apr 1 2013;85(7):898-912.

46. Woo KJ, Lim JH, Suh SI, *et al.* Differential inhibitory effects of baicalein and baicalin on LPS-induced cyclooxygenase-2 expression through inhibition of C/EBPbeta DNA-binding activity. *Immunobiology.* 2006;211(5):359–368.

47. Mahat MY, Kulkarni NM, Vishwakarma SL, *et al.* Modulation of the cyclooxygenase pathway via inhibition of nitric oxide production contributes to the anti-inflammatory activity of kaempferol. *Eur J Pharmacol.* Sep 2010; 642(1–3):169–176.

48. Brenner DR, Scherer D, Muir K, *et al.* A review of the application of inflammatory biomarkers in epidemiologic cancer research. *Cancer Epidemiol Biomarkers Prevent.* Sep 2014;23(9):1729–1751.

49. Peluso I, Miglio C, Morabito G, Ioannone F, Serafini M. Flavonoids and immune function in human: Asystematic review. *Crit Rev Food Sci Nutr.* 2015;55(3):383–395.

50. Bobe G, Murphy G, Albert PS, *et al.* Serum cytokine concentrations, flavonol intake and colorectal adenoma recurrence in the Polyp Prevention Trial. *Br J Cancer.* Oct 2010;103(9):1453–1461.

51. Mentor-Marcel RA, Bobe G, Sardo C, *et al.* Plasma cytokines as potential response indicators to dietary freeze-dried black raspberries in colorectal cancer patients. *Nutr Cancer.* Aug 2012;64(6):820–825.

52. Kim ER, Chang DK. Colorectal cancer in inflammatory bowel disease: The risk, pathogenesis, prevention and diagnosis. *World Journal of Gastroenterology.* Aug 2014;20(29):9872–9881.

53. Abraham C, Cho JH. Inflammatory bowel disease. *N Engl J Med.* Nov 2009;361(21):2066–2078.

54. Mayer L. Evolving paradigms in the pathogenesis of IBD. *J Gastroenterol.* Dec 2009.

55. Strober W, Fuss IJ. Proinflammatory cytokines in the pathogenesis of inflammatory bowel diseases. *Gastroenterology.* May 2011;140(6):1756–1767.

56. Molodecky NA, Soon IS, Rabi DM, *et al.* Increasing incidence and prevalence of the inflammatory bowel diseases with time, based on systematic review. *Gastroenterology.* Jan 2012;142(1):46–54 e42; quiz e30.

57. Thia KT, Loftus Jr. EV, Sandborn WJ, Yang SK. An update on the epidemiology of inflammatory bowel disease in Asia. *Am J Gastroenterol.* Dec 2008;103(12):3167–3182.

58. Ng SC, Woodrow S, Patel N, Subhani J, Harbord M. Role of genetic and environmental factors in British twins with inflammatory bowel disease. *Inflamm Bowel Dis.* May 2011.

59. Kaplan GG, Hubbard J, Korzenik J, *et al.* The inflammatory bowel diseases and ambient air pollution: A novel association. *Am J Gastroenterol.* Nov 2011;105(11):2412–2419.

60. Ananthakrishnan AN, McGinley EL, Binion DG, Saeian K. A nationwide analysis of changes in severity and outcomes of inflammatory bowel disease hospitalizations. *J Gastrointest Surg.* Feb 2011;15(2):267–276.

61. Henderson P, van Limbergen JE, Schwarze J, Wilson DC. Function of the intestinal epithelium and its dysregulation in inflammatory bowel disease. *Inflamm Bowel Dis.* Jan 2011;17(1):382–395.

62. Khor B, Gardet A, Xavier RJ. Genetics and pathogenesis of inflammatory bowel disease. *Nature.* Jun 2011;474(7351):307–317.

63. Bates J, Diehl L. Dendritic cells in IBD pathogenesis: An area of therapeutic opportunity? *J Pathol.* Jan 2014;232(2):112–120.

64. Danese S. Immune and nonimmune components orchestrate the pathogenesis of inflammatory bowel disease. *Am J Physiol. Gastrointest Liver Physiol.* May 2011;300(5):G716–722.

65. Zhang YZ, Li YY. Inflammatory bowel disease: Pathogenesis. *World Journal of Gastroenterology.* Jan 2014;20(1):91–99.

66. Korn T, Bettelli E, Oukka M, Kuchroo VK. IL-17 and Th17 Cells. *Annu Rev Immunol.* 2009;27:485–517.

67. Zenewicz LA, Flavell RA. Recent advances in IL-22 biology. *Int Immunol.* Mar 2011;23(3):159–163.

68. American Gastroenterological Association Institute. American Gastroenterological Association Institute medical position statement on the management of gastric subepithelial masses. *Gastroenterology.* Jun 2006; 130(7):2215–2216.

69. Bernstein CN. Treatment of IBD: Where we are and where we are going. *Am J Gastroenterol.* Jan 2015;110(1):114–126.

70. Leiman DA, Lichtenstein GR. Therapy of inflammatory bowel disease: What to expect in the next decade. *Curr Opin Gastroenterol.* Jul 2014; 30(4):385–390.

71. Creed TJ, Probert CS. Review article: Steroid resistance in inflammatory bowel disease — mechanisms and therapeutic strategies. *Alimentary Pharmacology and Therapeutics.* Jan 2007;25(2):111–122.

72. Thomas A, Lodhia N. Advanced therapy for inflammatory bowel disease: Aguide for the primary care physician. *J AmBoard Fam Med.* May–Jun 2014;27(3):411–420.

73. Swaminath A, Taunk R, Lawlor G. Use of methotrexate in inflammatory bowel disease in 2014: A user's guide. *World Journal of Gastrointestinal Pharmacology and Therapeutics.* Aug 2014;5(3):113–121.

74. Muller MR, Rao A. NFAT, immunity and cancer: Atranscription factor comes of age. *Nature Reviews:Immunology.* Sep 2010;10(9):645–656.

75. Valatas V, Vakas M, Kolios G. The value of experimental models of colitis in predicting efficacy of biological therapies for inflammatory bowel diseases. *Am J Physiol. Gastrointest Liver Physiol.* Dec 2013;305(11):G763–785.

76. DeVoss J, Diehl L. Murine models of inflammatory bowel disease (IBD): Challenges of modeling human disease. *Toxicol Pathol.* Jan 2014;42(1): 99–110.

77. Neurath MF. Animal models of inflammatory bowel diseases: Illuminating the pathogenesis of colitis, ileitis and cancer. *Dig Dis.* 2012;30 Suppl 1:91–94.

78. Randhawa PK, Singh K, Singh N, Jaggi AS. A review on chemical-induced inflammatory bowel disease models in rodents. *Korean Journal of Physiology and Pharmacology* (official journal of the Korean Physiological Society and the Korean Society of Pharmacology). Aug 2014;18(4):279–288.

79. Oh SY, Cho KA, Kang JL, Kim KH, Woo SY. Comparison of experimental mouse models of inflammatory bowel disease. *Int JMol Med.* Feb 2014; 33(2):333–340.

80. Ostanin DV, Bao J, Koboziev I, *et al.* T cell transfer model of chronic colitis: Concepts, considerations, and tricks of the trade. *Am J Physiol. Gastrointest Liver Physiol.*Feb 2009;296(2):G135–146.

81. Verdu EF, Deng Y, Bercik P, Collins SM. Modulatory effects of estrogen in two murine models of experimental colitis. *Am J Physiol. Gastrointest Liver Physiol.*Jul 2002;283(1):G27–36.

82. Mahler M, Bristol IJ, Leiter EH, *et al.* Differential susceptibility of inbred mouse strains to dextran sulfate sodium-induced colitis. *Am J Physiol.* Mar 1998;274(3 Pt 1):G544–551.

83. Wullaert A, Bonnet MC, Pasparakis M. NF-kappaB in the regulation of epithelial homeostasis and inflammation. *Cell Res.* Jan 2011;21(1):146–158.

84. Visekruna A, Volkov A, Steinhoff U. A key role for NF-kappaB transcription factor c-Rel in T-lymphocyte-differentiation and effector functions. *Clinical and Developmental Immunology.* 2012;2012: Article ID 239368.

85. Feng J, Guo C, Zhu Y, *et al.* Baicalin down regulates the expression of TLR4 and NFkB-p65 in colon tissue in mice with colitis induced by dextran sulfate sodium. *International Journal of Clinical and Experimental Medicine.* 2014;7(11):4063–4072.

86. Mascaraque C, Gonzalez R, Suarez MD, Zarzuelo A, Sanchez de Medina F, Martinez-Augustin O. Intestinal anti-inflammatory activity of apigenin K in two rat colitis models induced by trinitrobenzenesulfonic acid and dextran sulphate sodium. *Br J Nutr.* Feb 2015;113(4):618–626.

87. Cui L, Feng L, Zhang ZH, Jia XB. The anti-inflammation effect of baicalin on experimental colitis through inhibiting TLR4/NF-kappaB pathway activation. *Int Immunopharmacol.* Nov 2014;23(1):294–303.

88. Dou W, Mukherjee S, Li H, *et al.* Alleviation of gut inflammation by Cdx2/Pxr pathway in a mouse model of chemical colitis. *PLoS One.* 2012; 7(7):e36075.

89. Nishitani Y, Yamamoto K, Yoshida M, *et al.* Intestinal anti-inflammatory activity of luteolin: Role of the aglycone in NF-kappaB inactivation in macrophages co-cultured with intestinal epithelial cells. *BioFactors.* Sep–Oct 2013;39(5):522–533.

90. Karrasch T, Kim JS, Jang BI, Jobin C. The flavonoid luteolin worsens chemical-induced colitis in NF-kappaB(EGFP) transgenic mice through blockade of NF-kappaB-dependent protective molecules. *PLoS One.* 2007;2(7):e596.

91. Min JK, Lee CH, Jang SE, *et al.* Liquiritigenin ameliorates TNBS-induced colitis in mice. *J Gastroenterol Hepatol.* Oct 2014.

92. Al-Rejaie SS, Abuohashish HM, Al-Enazi MM, Al-Assaf AH, Parmar MY, Ahmed MM. Protective effect of naringenin on acetic acid-induced ulcerative colitis in rats. *World Journal of Gastroenterology.* Sep 2013;19(34):5633–5644.

93. Azuma T, Shigeshiro M, Kodama M, Tanabe S, Suzuki T. Supplemental naringenin prevents intestinal barrier defects and inflammation in colitic mice. *J Nutr.* Jun 2013;143(6):827–834.

94. Dou W, Zhang J, Sun A, *et al*. Protective effect of naringenin against experimental colitis via suppression of Toll-like receptor 4/NF-kappaB signalling. *Br J Nutr.* Aug 2013;110(4):599–608.

95. Sanchez de Medina F, Galvez J, Romero JA, Zarzuelo A. Effect of quercitrin on acute and chronic experimental colitis in the rat. *J Pharmacol Exp Ther.* Aug 1996;278(2):771–779.

96. Sanchez de Medina F, Vera B, Galvez J, Zarzuelo A. Effect of quercitrin on the early stages of hapten induced colonic inflammation in the rat. *Life Sciences.* May 2002;70(26):3097–3108.

97. Comalada M, Camuesco D, Sierra S, *et al*. In vivo quercitrin anti-inflammatory effect involves release of quercetin, which inhibits inflammation through down-regulation of the NF-kappaB pathway. *Eur J Immunol.* Feb 2005; 35(2):584–592.

98. Kwon KH, Murakami A, Tanaka T, Ohigashi H. Dietary rutin, but not its aglycone quercetin, ameliorates dextran sulfate sodium-induced experimental colitis in mice: Attenuation of pro-inflammatory gene expression. *Biochem Pharmacol.* Feb 2005;69(3):395–406.

99. Camuesco D, Comalada M, Rodriguez-Cabezas ME, *et al*. The intestinal anti-inflammatory effect of quercitrin is associated with an inhibition in iNOS expression. *Br J Pharmacol.* Dec 2004;143(7):908–918.

100. Sotnikova R, Nosalova V, Navarova J. Efficacy of quercetin derivatives in prevention of ulcerative colitis in rats. *Interdiscipl Toxicol.* Mar 2013;6(1):9–12.

101. Guazelli CF, Fattori V, Colombo BB, *et al*. Quercetin-loaded microcapsules ameliorate experimental colitis in mice by anti-inflammatory and antioxidant mechanisms. *J Nat Prod.* Feb 2013;76(2):200–208.

102. Castangia I, Nacher A, Caddeo C, *et al*. Therapeutic efficacy of quercetin enzyme-responsive nanovesicles for the treatment of experimental colitis in rats. *Acta Biomaterialia.* Feb 2015;13:216–227.

103. Kim H, Kong H, Choi B, *et al*. Metabolic and pharmacological properties of rutin, a dietary quercetin glycoside, for treatment of inflammatory bowel disease. *Pharm Res.* Sep 2005;22(9):1499–1509.

104. Mascaraque C, Aranda C, Ocon B, *et al*. Rutin has intestinal antiinflammatory effects in the CD4+ CD62L+ T cell transfer model of colitis. *Pharmacological Research* (the official journal of the Italian Pharmacological Society). Dec 2014;90:48–57.

105. Sakai T, Furoku S, Nakamoto M, *et al.* Soy isoflavone equol perpetuates dextran sulfate sodium-induced acute colitis in mice. *Biosci Biotechnol Biochem.* 2011;75(3):593–595.

106. Seibel J, Molzberger AF, Hertrampf T, Laudenbach-Leschowski U, Degen GH, Diel P. In utero and postnatal exposure to a phytoestrogen-enriched diet increases parameters of acute inflammation in a rat model of TNBS-induced colitis. *Arch Toxicol.* Dec 2008;82(12):941–950.

107. Sadowska-Krowicka H, Mannick EE, Oliver PD, *et al.* Genistein and gut inflammation: Role of nitric oxide. *Proceedings of the Society for Experimental Biology and Medicine. Society for Experimental Biology and Medicine.* Mar 1998;217(3):351–357.

108. Seibel J, Molzberger AF, Hertrampf T, Laudenbach-Leschowski U, Diel P. Oral treatment with genistein reduces the expression of molecular and biochemical markers of inflammation in a rat model of chronic TNBS-induced colitis. *Eur J Nutr.* Jun 2009;48(4):213–220.

109. Sato K, Kanazawa A, Ota N, Nakamura T, Fujimoto K. Dietary supplementation of catechins and alpha-tocopherol accelerates the healing of trinitrobenzene sulfonic acid-induced ulcerative colitis in rats. *J Nutr Sci Vitaminol.* Dec 1998;44(6):769-778.

110. Abboud PA, Hake PW, Burroughs TJ, *et al.* Therapeutic effect of epigallo-catechin-3-gallate in a mouse model of colitis. *Eur J Pharmacol.* Jan 2008; 579(1–3):411–417.

111. Maity S, Ukil A, Karmakar S, *et al.* Thearubigin, the major polyphenol of black tea, ameliorates mucosal injury in trinitrobenzene sulfonic acid-induced colitis. *Eur J Pharmacol.* May 2003;470(1–2):103–112.

112. Ukil A, Maity S, Das PK. Protection from experimental colitis by theaflavin-3,3'-digallate correlates with inhibition of IKK and NF-kappaB activation. *Br J Pharmacol.* Sep 2006;149(1):121–131.

113. Mochizuki M, Hasegawa N. (-)-Epigallocatechin-3-gallate reduces experimental colon injury in rats by regulating macrophage and mast cell. *Phytother Res.* Jan 2010;24 Suppl 1:S120–122.

114. Chiou YS, Ma NJ, Sang S, Ho CT, Wang YJ, Pan MH. Peracetylated (-)-epigallocatechin-3-gallate (AcEGCG) potently suppresses dextran sulfate sodium-induced colitis and colon tumorigenesis in mice. *J Agric Food Chem.* Apr 2012;60(13):3441–3451.

115. Ran ZH, Chen C, Xiao SD. Epigallocatechin-3-gallate ameliorates rats colitis induced by acetic acid. *Biomed Pharmacother.* Mar 2008;62(3):189–196.

116. Dryden GW, Lam A, Beatty K, Qazzaz HH, McClain CJ. A pilot study to evaluate the safety and efficacy of an oral dose of (-)-epigallocatechin-3-gallate-rich polyphenon E in patients with mild to moderate ulcerative colitis. *Inflamm Bowel Dis.* Aug 2013;19(9):1904–1912.

117. Guan F, Liu AB, Li G, *et al.* Deleterious effects of high concentrations of (−)-epigallocatechin-3-gallate and atorvastatin in mice with colon inflammation. *Nutr Cancer.* Aug 2012;64(6):847–855.

118. Yum HW, Zhong X, Park J, *et al.* Oligonol inhibits dextran sulfate sodium-induced colitis and colonic adenoma formation in mice. *Antioxid Redox Signal.* Jul 2013;19(2):102–114.

Flavonoids and Cancers of the Gastrointestinal Tract*

4

Overview of gastrointestinal cancers. Gastrointestinal tract (GI) cancers include esophageal (squamous cell carcinoma and adenocarcinoma), stomach (adenocarcinoma), liver (hepatocellular carcinoma), gastrointestinal stromal, pancreatic, and colorectal cancers.[1] The majority of GI cancers are closely linked to inflammation and harbor activating mutations of the RAS oncogene. In many cases, such as Barrett's esophagus and colorectal cancer, neoplastic development is preceded by localized inflammation. However, it has also been noted that the "gene responsible for stimulating proliferation or restricting growth in one tissue may not have the same function in another tissue."[2] Thus, the tissue microenvironment

*****Abbreviations**: AID, activation-induced cytidine deaminase; AKT, v-akt murine thymoma viral oncogene homolog 1; AOM, azoxymethane; APC, adenomatous polyposis coli; BAX, BCL2-associated X protein; BMI, body mass index; BRAF, B-Raf proto-oncogene, serine/threonine kinase; COX-2, cyclooxygenase 2; DMH, 1,2-dimethylhydrazine; DR3/4/5, death receptor 3/4/5; EGCG, (–)-epigallocatechin-3-gallate; EGFR, epidermal growth factor receptor; ERBB2, erb-b2 receptor tyrosine kinase 2; ERK, extracellular signal-regulated kinase; FOXM1, forkhead box M1; GI, gastrointestinal; HER2, human epidermal growth factor 2; IL1β/6/8/10, interleukins 1β/6/8/10; iNOS, inducible nitric oxide synthase; KIT, v-kit Hardy-Zuckerman 4 feline sarcoma viral oncogene homolog; MAPK, mitogen activated protein kinase; MAPK8/JNK, mitogen-activated protein kinase 8; MET, hepatocyte growth factor receptor; mTOR, mechanistic target of rapamycin; NSAID, non-steroidal anti-inflammatory drug; NFκB, nuclear factor of kappa light polypeptide gene enhancer in B-cells 1; PIK3CA, phosphatidylinositol-4,5-bisphosphate 3-kinase, catalytic subunit alpha; PTEN, phosphatase and tensin homolog; RAS, rat sarcoma viral oncogene homolog 1; RONS, reactive oxygen and reactive nitrogen species; ROS, reactive oxygen species; STAT3, signal transducer and activator of transcription 3; TLR, toll-like receptor; TNF, tumor necrosis factor; TNFα, tumor necrosis factor alpha; TP53, tumor protein 53; TRAIL, tumor necrosis factor (ligand) superfamily, member 10 (officially known as TNFSF10); WNT, wingless-type MMTV integration site family.

105

likely plays an important role, not only in tumor etiology but also in the response of the tumor cell to the administered therapeutic agent. These issues will be considered as we examine the pathogenesis of GI cancers, some of the chemopreventive approaches being considered to inhibit GI cancers, and the potential of flavonoids as chemopreventive agents.

Incidence and etiology of gastrointestinal cancers. The highest rates of esophageal cancer are found in South African males where an incidence rate of 22.3 per 100,000 is observed.[1,3] However, high incidences are also observed in Asia, in particular, China and Iran. In these countries, squamous cell carcinomas are the most common and risk factors include poor nutritional status, low intake of fruits and vegetables, and drinking beverages at high temperatures. In the United States and other more developed countries, adenocarcinomas of the esophagus are more common and are rising presumably due to obesity and "over-nutrition." The majority of primary esophageal adenocarcinomas develop in patients with Barrett's esophagus, a condition of prolonged reflux disease and chronic esophageal inflammation. Barrett's esophagus is characterized as an abnormal transformation that occurs in the distal esophagus. Here, the squamous cell epithelium becomes metaplastic and assumes a column-like appearance. The transformation process is thought to arise from the cellular damage inflicted by high exposure to stomach acid and bile salts as well as the reflux of pancreatic enzymes. As a result, an inflammatory process ensues that typically becomes unresolved. High expression levels of COX-2 are often observed in esophageal adenocarcinomas. Comparison of the genetic landscape of esophageal squamous cell carcinomas obtained by exome sequencing of normal tissue versus tumor tissue has recently revealed that 99% of the tumor samples contain mutations in genes involved in cell cycle and apoptosis.[4] The most frequently altered gene found in 93% of the samples was TP53.

High incidences of stomach cancer are found in Eastern Asia, Eastern Europe, and South America, and are associated with the prevalence of *Helicobacter pylori* (*H. pylori*) infection, smoking, poor availability of fresh fruits and vegetables, and high intake of salted and preserved foods.[1,5] Worldwide, the infection rate of *H. pylori* is relatively high; approximately 50% of the total population is infected. Further, 80% of gastric cancers are thought to arise as a consequence of *H. pylori* infection. The events involved in the development of *H. pylori*-associated gastric cancer initiates with oxidative stress, formation of reactive oxygen and nitrogen species, and damage to endogenous DNA. In addition, DNA repair pathways are inhibited and NFκB pro-inflammatory and WNT/β-catenin pathways are upregulated. High salt diets are thought to exacerbate these conditions by enhancing the colonization of *H. pylori* and contributing to hypergastrinemia. Additional molecular mechanisms involved in the pathogenesis of stomach cancer include overexpression of tyrosine kinase receptors, in particular ERBB2/HER2 and MET as well as constitutive activation of the downstream signaling pathways MAPK and PIK3CA/AKT/mTOR.[6]

Countries with the highest incidences of hepatocellular carcinoma are China, Middle Africa, Japan, and Eastern Africa.[7,8] Patients at the highest risk of developing hepatocellular carcinoma are those chronically infected with hepatitis B, hepatitis C, or who have developed liver cirrhosis. Other contributor risk factors, particularly in the United States, are diabetes and nonalcoholic fatty liver disease. Approximately 80% of hepatocellular carcinomas occur as a consequence of infection with either hepatitis B, hepatitis C, or both. Like many GI cancers, hepatocellular carcinoma is well characterized as an inflammation-associated cancer wherein a condition of chronic inflammation is observed throughout the progression of the disease. In addition to the inflammatory response mediated by the host's immune system, the virally encoded proteins alter the ability of the host to express genes involved in proliferation, cell survival, apoptosis, angiogenesis, and invasion and metastasis. Chronic infection ultimately results in a poor innate response that is insufficient for mounting an appropriate adaptive immune response. As the carcinogenic disease progresses, aberrant activity of the EGFR/RAS/PIK3CA/mTOR, WNT/Hedgehog, MET, and apoptotic signaling pathways are also observed.[6]

Gastrointestinal stromal tumors are relatively rare and are thought to originate in the muscularis propria.[6] They occur largely in the stomach, but are also found in the jejunum, ileum, duodenum, rectum, and esophagus. Gastrointestinal tumors appear to arise primarily due to a hereditary mutation in the KIT oncogene. Due to the lack of published studies that address the impact of flavonoids on gastrointestinal stromal tumorigenesis, it will not be subject to further discussion.

Pancreatic cancer is the 11th most common cancer worldwide and is more prevalent in more developed countries.[9–11] Risk factors include genetics, diabetes, infection with *H. pylori*, and a history of pancreatitis. Mutations in genes involved in proteolysis are amongst those that predispose an individual to developing pancreatic cancer. This includes mutations in PRSS1/2 (protease, serine 1/2), which are members of the trypsin family of serine proteases, SPINK1 (serine protease inhibitor Kazal-type 1), a trypsin inhibitor, and CTRC (chymotrypsin C), a member of the peptidase S1 family that harbors chemotrypsin-like protease activity.

The most common pancreatic neoplasia, pancreatic ductal adenocarcinoma, is thought to originate in the acinar cells and involve progression from acinar to ductal metaplasia. The majority of the pancreatic tissue is composed of the exocrine compartment that contains the endocrine islets. The primary function of the exocrine compartment is to synthesize and transport the enzymes required for intestinal digestion. Inflammation in the pancreas initially occurs as acute pancreatitis during which the acinar secretory enzymes are inappropriately activated which ultimately damages the pancreatic tissue. Repeated episodes of pancreatitis result in unresolved inflammation, fibrosis, and formation of abnormal ducts. While pancreatitis is the strongest risk factor for developing pancreatic cancer, less than 5% of patients with chronic pancreatitis will develop pancreatic cancer. While the aforementioned mutations increase disease susceptibility, mutations involved in oncogene activation (i.e., KRAS) and inactivation of tumor suppressor proteins (i.e., TP53) are typically involved in tumor progression. In fact, over 90% of pancreatic tumors harbor mutations associated with RAS activation.[6] The major involved pro-inflammatory factors include COX-2, NFκB, and STAT3.[11]

The incidence of colorectal cancer, the third most common cancer worldwide, is highest in Australia, New Zealand, Europe, and North America.[1] The incidence is rapidly increasing in countries such as Spain and within Eastern Asia and Eastern Europe, presumably due to the increased prevalence of risk factors such as poor diet, obesity, and cigarette smoking. Additional risk factors of colorectal cancer include genetics, physical inactivity, consumption of red and processed meats, excessive alcohol consumption, and inflammatory bowel disease (IBD). Of these risk factors, the most important appear to be genetics and the chronic inflammation that accompanies IBD. Colorectal cancers are primarily found in the descending colon and the rectum and are thought to arise from adenomatous polyps.[6,12] With the polyps, a series of changes often occur which involve the acquisition of KRAS mutations (leading to constitutive activation of the MAPK pathway) and inactivation of tumor suppressor genes such as APC. However, like many other cancers, colorectal cancer is highly heterogeneous and includes sporadic and colitis-associated colorectal cancers.[13] The heterogeneity of colorectal cancer is thought to arise from multiple events including the order in which the tumor cell acquires its mutations, its cell type of origin, and the absence or presence of chronic inflammation as well as the risk factors described above.

Three main subtypes of sporadic colorectal cancer have been proposed based on molecular and clinical attributes. By understanding the nature of the subtypes, we can better predict how patients will respond to cancer treatment and perhaps chemopreventive strategies. The first subtype is characterized by mutations in KRAS and TP53, high activity of the WNT signaling pathway, and significant chromosomal instability. This subtype appears to originate primarily from the epithelial (i.e., enterocytes) cells. The second subtype harbors a phenotype that is characterized by the presence of microsatellite instability and methylated CpG islands. These tumors are frequently highly infiltrated with immune cells and are often found within the ascending colon. The third subtype is characterized by tumors with microsatellite and chromosomal instability and are also enriched for mutations in BRAF and PIK3CA. These bear a mesenchymal phenotype. Patients with Lynch syndrome, for example, are amongst those with the highest risk for developing hereditary colorectal cancer.[14] Lynch syndrome predisposes an individual to a number of cancers with colorectal and endometrial being the most common. The underlying cause is autosomal dominate heterozygous germline mutations in key genes involved in DNA repair. As a result of defective DNA repair, microsatellite instability emerges.

Colitis-associated cancer frequently develops in patients with IBD with incidences that are higher in patients with ulcerative colitis as compared to those with Crohn's disease.[15–17] The risk of developing colitis-associated cancer is associated with the duration of the disease and appears to be linked to the development of primary sclerosing cholangitis and is typically higher in patients with severe and persistent forms of IBD. Primary sclerosing cholangitis occurs most frequently in male IBD patients and is characterized by hepatitis, liver damage, and loss of bile ducts that is a consequence of progressive liver fibrosis.[18] It is thought that an autoimmune event may contribute to the etiology of the disease state. The progression of colitis-associated cancer occurs via increasing grades of dysplasia, which is initially observed as low grade and advances to high grade dysplasia and finally, adenocarcinomas.[17] The dysplastic lesions may have a flat or raised appearance. Histologically, they are viewed as distorted crypts. Within low grade dysplasia, the polarization of the epithelial cells is maintained. However, within high grade dysplasia, polarization is lost and the cells acquire a high nuclear to cytoplasmic ratio. With respect to mutational events that occur in the progression of colitis-associated cancer, key and early events are mutations that abrogate the function of TP53.[15–17] TP53 mutations are observed in 50–80% of patients with colitis-associated cancers and are often observed prior to the formation of observable dysplastic tissue. Another common event detected in the inflamed tissue of a patient with IBD is DNA hypermethylation. For example, the tumor suppressor and cell cycle inhibitor p16 is hypermethylated in up to 100% of the patients. As tumors form and progress, aneuploidy and loss of heterozygosity occur which are thought to contribute to the loss of function of both TP53 and APC as well as the formation of additional mutations. Additional events include oxidative stress-induced telomere shortening that is thought to contribute to genetic instability. Mutations in genes involved in DNA repair and in KRAS are often detected in the tissues with high grade dysplasia and within the colon tumors.

Inflammation and GI cancers. GI and, in particular, colorectal cancers represent an ideal paradigm for studying the role of inflammation in tumorigenesis as the involved mucosal tissue is constantly exposed to agents, such as microbes, that initiate inflammatory events, and the resident epithelial cells are amongst the few within the adult that undergo constant proliferation.[15,17,19,20] In fact, it is thought that this state of perpetual epithelial cell proliferation is a key event involved in the transition from IBD to colitis-associated cancer. The link between inflammation and GI cancers can be observed not only in colitis-associated cancers, but also in patients with impaired immune responses. For example, patients harboring the AIDS virus are at high risk for developing neoplasias of the colon as well as pancreas, lung, kidney, head and neck, and melanomas. As previously mentioned, other infectious agents that increase an individual's susceptibility to developing GI tract cancers are hepatitis B and hepatitis C viruses and *H. pylori*. Finally, links between GI tract cancer and inflammation can be observed in successful chemopreventive anti-inflammatory-based approaches. For example, epidemiological evidence indicates that the administration of non-steroidal anti-inflammatory drugs such as aspirin is associated with a decreased risk of developing GI tract cancers. Some evidence also indicates that the use of mesalazine and aminosalicylates may also confer a protective effect. Finally, the extent to which patients with IBD are at risk for developing cancer appears to be related to how well their inflammation is controlled and the extent of mucosal healing that occurs.

It is now thought that inflammation occurs at every step of the carcinogenic pathway.[17] The role of inflammation in the very early steps of tumorigenesis involving the progenitor stem cells is best exemplified in colorectal cancer. The epithelial stem cells located at the bottom of the intestinal crypts are the most likely originators of colorectal tumors. These cells play an important role in the maintenance of the intestinal epithelial barrier. At the bottom of the crypts, the young, transit amplifying cells proliferate as they migrate up toward the surface and intestinal villi. During their journey, they also progressively differentiate. While the small intestine has both crypts and villi, the large intestine has only crypts. Regardless, both are lined with transit amplifying cells that differentiate into mucosal enterocytes, enteroendocrine cells, goblet, and paneth cells. While the exact identity of the intestinal stem cell is still under discussion, one idea that is gaining popularity is that two populations of intestinal stem cells exist. One of these is an actively dividing population that harbors high regenerative capacity and can quickly respond to injury or stress signals. The other population is thought of as the "reserve" population that is largely quiescent within the normal tissue.

Damage to epithelial stem cells is likely imposed by persistent oxidative stress generated by the innate and adaptive immune cells.[15,17,19] At the initial stage of tumorigenesis and in the presence of chronic inflammation, excess formation of RONS damages DNA, yielding mutations in sensors of DNA damage and tumor suppressor proteins such as TP53. In addition, the inflammatory conditions are pro-proliferative and thus stimulate hyperproliferation of the normal epithelial cells, thereby enhancing their risk for undergoing neoplastic transformation. Involved growth and survival stimuli include IL1β, TNFα, IL6 and the WNT/β-catenin, AKT, NFκB, and STAT3 pathways. These, in addition to the upregulated enzymes prostaglandin G/H synthase and COX-2, are thought to be the most important mediators of inflammation-induced GI cancer.

The importance of NFκB in GI tumorigenesis is indicated by observations that in patients with *H. pylori*-associated gastritis, hepatitis C-associated chronic liver disease, and IBD, a dysregulated (i.e., constitutively activated) NFκB pathway is a common feature. Further, links between the NFκB pathway and DNA mutagenesis are emerging. For example, an enzyme called AID, activation-induced cytidine deaminase, is expressed only in activated B cells and is regulated by NFκB and following TNFα stimulation. Its function is to induce genetic mutations. It accomplishes this task by deaminating cytosine residues of DNA which ultimately converts a C:G to a T:A base pairing. Infection with *H. pylori* leads to NFκB activation, induction of AID, and ultimately, formation of mutations in TP53. Further, in the IL10$^{-/-}$ mouse model of experimental colitis, colitis-associated tumors do not develop in mice that lack expression of AID.

Upstream regulators of the NFκB pathway, such as the toll-like receptors (TLRs) that sense the presence of microbes, are also implicated in a number of GI cancers including colorectal, esophageal, gastric, and hepatocellular carcinomas.[21–24] The first indications that an association between the TLRs and colorectal cancer existed arose from the observations that mice lacking microbial colonization failed to develop colorectal tumors. In human patients, polymorphisms in TLR4 are linked to increased susceptibility to developing colorectal cancer and expression levels of TLR4 were often increased in colorectal tumors. Finally, mice lacking expression of TLR4 failed to develop colitis-induced colorectal tumors. It is not yet clear whether TLR activation that contributes to colo-rectal cancer originates from microbial or host sources.

TLRs are also involved in the pathogenesis of other GI cancers. With respect to the esophagus, TLRs 1–10 are expressed in normal esophageal tissues and increased expression levels and/or activity of TLR 3, 4, 5 and 9 are linked to the progression to esophageal cancer.[22] As previously mentioned, *H. pylori* infection is associated with the development of stomach cancer.[23] A number of TLRs, in particular, TLR2, 4, and 5 are involved in the host inflammatory response to *H. pylori* and may also be involved in the chronic inflammation that accompanies stomach tumor progression. Further, polymorphisms in TLR signaling are implicated in determining an individual's susceptibility to the gastric cancer pathogenesis. Finally, with respect to hepatocellular carcinomas, expression levels of TLR7 and TLR9 are increased in tissues obtained from hepatocellular carcinomas as compared to tissue from normal, cirrhosis, or viral hepatitis.[25]

The IL6/STAT3 pathway is also a well-recognized player in inflammation associated with GI cancer. IL6 is produced primarily by innate immune cells, in particular, the monocytes and macrophages. IL6 activates both the JAK/STAT and PIK3CA/AKT signaling pathways which culminate in the activation of STAT3. STAT3 confers oncogenic properties as it regulates genes involved in proliferation, cell survival, angiogenesis, and inflammation. In addition, STAT3 activation can dampen the anti-tumor immune response.[26] Further, it promotes stem cell self-renewal and facilitates gene methylation and silencing. This latter action is thought to involve its interactions with DNA methyltransferase and histone deacetylase. A major target of these events is TP53. Depending on the genetics of the host (i.e., absence or presence of PTEN expression), STAT3 can also act as a tumor suppressor.

It is now recognized that considerable crosstalk between the TLR/NFκB and IL6/STAT3 pathways exist.[21,27] They often regulate an overlapping set of genes and appear to synergistically promote tumorigenesis. IL6 family members have been shown to modulate TLR signaling in a context-dependent manner. For example, IL11-induced STAT3 in the gastric epithelium transcriptionally upregulates TLR2, but not TLR4. In vastus-latealis tissue, however, IL6 upregulation of STAT3 results in increased TLR4 expression. Finally, another IL6 family member, IL27, activates both NFκB and STAT3 resulting in increased expression of TLR4. Understanding these crosstalk mechanisms and consequences may be key to our understanding of how chronic inflammation contributes to tumorigenesis and how these pathways may be effectively targeted by appropriate chemopreventive agents.

The microbiome and GI cancer. The interplay between microorganisms and cancer has historically been focused on the actions of known pathogens such as *H. pylori*, hepatitis B, and hepatitis C, and their direct role in tumor initiation and progression.[28] While it is well understood that the presence of these agents initiates conditions of sustained inflammation within the infected tissue, their impact on the commensal microorganisms that typically reside within healthy tissue has not been well characterized until relatively recently. It is now thought that microbial dysbiosis likely plays an important role in the pathogenesis of many cancers including that of the colon, stomach, esophagus, pancreas, larynx, and gallbladder. For example, it has become apparent that infection with *H. pylori* not only initiates a host immune response and garners control of the host's cellular control system, but also disrupts the host's gut microbial community. Similarly, disruptions in the microbial community have been observed in patients with Barrett's esophagus where the microbial population shifts from a dominance of gram positive microorganisms to one composed primarily of gram negative microorganisms. Similar shifts have been observed in patients with pancreatic cancer; however, the observed changes were with respect to their oral microbial communities. Given that the current literature is more detailed and extensively focused on the gut microbiome and colorectal cancer versus other GI cancers, we will limit further discussion to this topic.

Within the colon, approximately 90% of the cell population is represented by microbial organisms.[29–31] In fact, it has been reported that within the human gut, the bacteria cells exceed that of human cells by ten-fold in number and harbor a genome with 150-fold more genes than that of the human cells. The major bacterial species that typically reside in the colon of a healthy individual are the gram negative *Bacteroidetes* and gram positive *Firmicytes*. Minor species include *Proteobacteria*, *Actinobacteria*, and *Fusobacteria*. The colon microbiota is formed within the first two years of life and although is relatively stable thereafter, it varies considerably from one individual to the other. It is thought to play integral roles in the development of the immune system, prevention of the colonization of pathogens, maintenance of barrier function, production of micronutrients such as vitamins B_{12} and K, and metabolism of drugs, xenobiotics, and dietary components. Major contributors to the wide interindividual variation in the composition of the colon microbiota include the genetic polymorphisms of the host, environmental factors, diet, and disease states, in particular those involving chronic inflammation such as obesity and IBD. The mucosal tissue of patients with chronic inflammatory conditions such as IBD is typically populated by a less diverse microbial population as compared to that of healthy individuals.

Within the large bowel, two bacterial communities exist. One resides within the lumen and the other within the mucosal epithelial cells of the crypts. In the mucosa, the bacterial cells are highly adherent which allows them to extensively interact with the host immune system. For this reason, it is thought that these bacterial cells may be of most importance when considering the role of the microbiome in colorectal cancer.

The role of the gut microbiome in metabolizing dietary substances likely plays a role in the development of colorectal cancers and the composition of the diet can impact the population of the resident microorganism.[32] For example, diets high in fiber typically result in an abundance of *Prevotrella* spp. whereas "western" diets high in fat and proteins favor formation of *Bacteroides* spp. Individuals with metabolic syndrome or who are obese tend to have a higher *Bacteroides* spp. to *Firmicutes* spp. ratio. The end products of bacterial metabolism are primarily gases and organic acids. Digestion of complex carbohydrates by the commensal bacteria results in the production of short chain fatty acids such as butyrate. Within the human colon, the major butyrate producers are *Faecalibacterium prausnitzii* and *Eubacterium/Roseburia* and butyrate is also generated by some species of *Firmicutes*. High colonic butyrate concentrations can have significant functional consequences. Butyrate is known to enhance gut motility and limit the growth of pathogenic microorganisms. Since it is the preferential energy source of the colonic epithelial cells, it can regulate their metabolic activity and proliferation. It is in general, thought of as "anti-tumorigenic" as it decreases the activity of histone deacetylases and in this manner, is thought to decrease the ability of the colonic macrophages to produce IL6 and other pro-inflammatory cytokines. In addition, it modifies T cell differentiation and favors formation of the "anti-inflammatory" T_{reg} subpopulation. It has also been described as harboring selective apoptosis activities. However, the properties of butyrate may be biphasic; at low concentrations they may be primarily pro-inflammatory and in this manner contribute to gut remodeling. At high concentrations, the activities of butyrate may be primarily anti-inflammatory.

Bacterial metabolism of ingested protein results in the formation of polyamines, hydrogen sulphide, and N-nitrosocompounds. These bacterial metabolites are thought to exert primarily detrimental effects. At low concentrations, polyamines are involved in normal physiological processes such as maintaining the structural integrity of membranes and nucleic acids. At high concentrations, however, polyamines are associated with cellular toxicity, presumably due to the oxidative stress associated with their catabolism. The generation of high levels of hydrogen sulphide by bacteria such as *Desulfovibrio* spp. is toxic to the colonic epithelial cells and thereby contributes to loss of the integrity of the intestinal barrier. In addition, it has been shown to exhibit genotoxic properties and in this manner contribute to DNA damage. N-nitrosocompounds thought to be generated by Proteobacteria are also mutagenic and capable of damaging the mucosal layer. Bacterial metabolism of fat and bile acids to secondary bile acids can also contribute to adverse events. Secondary bile acids of concern are deoxycholic acid and lithocholic acid which can be generated by microorganisms such as *Clostridium scindens*. The effects of these bile acids include the generation of ROS and activation of nuclear receptors that can modulate the apoptotic response.

The extent to which gut microorganisms may contribute to the development of colorectal cancers may depend on their relative virulence.[31,33] Highly virulent forms such as *H. pylori* or the hepatitis viruses are of sufficient virulence to directly contribute to the etiology of the disease. Those with lesser virulence, however, likely act in a more indirect manner by contributing to the overall dysbiotic composition of the microbial population. As mentioned above, the dietary end products generated by a number of microorganisms can contribute to the overall oxidative stress levels of the tissue. Additional microorganisms that have been found to exert adverse effects include *Enterococcus faecalis* and *Bacteriodes fragilis*. *Enterococcus faecalis* and *Bacteriodesfragilis* are capable of producing DNA damaging ROS and enhancing the NFκB/STAT3 pathways. In patients with chronic inflammatory conditions, changes in the microbiome often include increases in gram bacteria that produce lipopolysaccharides. These higher levels of lipopolysaccharides are detected by the toll-like receptors (i.e., TLR4) which then activate NFκB and enhance the production of pro-inflammatory cytokines. Patients with colorectal cancer are thought to harbor proportionally high levels of *Enterobacteriaceae, Bacteroides, Enterococcus, Fusobacterium* and *Streptococcus*. However, given the large inter-individual variations and uncertainty surrounding sampling procedures, it is currently difficult to obtain consistent results. It is also likely that as the disease state progresses, the environmental conditions within the colon changes, thereby allowing for opportunistic bacteria to outgrow the normal commensal resident bacteria.

Chemoprevention of esophageal cancer. Given the significant role of acid reflux and chronic inflammation in driving the events leading up to esophageal cancer, pharmacological approaches currently considered for use as chemopreventive agents are proton pump inhibitors (i.e., omeprazole), anti-inflammatory agents (aspirin and other NSAIDS), and statins.[34,35] NSAIDS are thought to exert their chemopreventive activities via both COX-2-dependent and COX-2-independent mechanisms.[36] COX-2-independent mechanisms include modulation of the NFκB, WNT/β-catenin, DNA mismatch repair, and cell cycle progression pathways. Despite their effectiveness, use of NSAIDS as chemopreventive agents is curtailed by safety concerns. For example, use of aspirin is associated with nausea and dyspepsia, formation of GI ulcers, increased risk of bleeding, and an increased risk of Reyes syndrome in the elderly.[37] Use of celecoxib is associated with significant cardiovascular toxicities.[38] However, its use may be problematic only in patients that have been identified as being at high risk for developing cardiovascular events. Statins are widely prescribed cholesterol-lowering agents with good safety profiles.[39] The most widely reported adverse effect is myopathy. The proposed underlying mechanisms include their inhibition of HMG-CoA reductase which ultimately inhibits the post-translational prenylation of RAS thereby affecting the RAS-MAPK-ERK pathway. By inhibiting this pathway, cell proliferation is attenuated and caspase-induced apoptosis is increased. In addition, statins block the proteosomal pathway which leads to sustained expression of key cell cycle inhibitors, p21 and p27.

Pre-clinical and epidemiological evidence has indicated that proton pump inhibitors may be effective in not only relieving the symptoms associated with Barrett's esophagus or GI reflux disease, but may also slow the progression to esophageal adenocarcinomas.[34,35] The proposed underlying mechanisms include inhibition of inflammation and proliferation. However, a concern is that proton pump inhibitors may also be capable of increasing cell proliferation via an increase in the expression levels of gastrin. A randomized clinical study involving 90 patients with Barrett's esophagus indicated that there is no evidence of long-term effects on markers of proliferation, apoptosis, and cell migration.[40] Further, a systematic review of 25 studies representing 1,696 patients with Barrett's esophagus reported that the incidences of esophageal adenocarcinomas appeared to be similar in patients treated with proton pump inhibitors versus those treated with surgical interventions.[41] Resolution on the issue of the chemopreventive effectiveness of proton pump inhibitors may be reached upon completion of an ongoing randomized trial called AspECT.[42] This trial also includes treatment with aspirin and thus may also lend insights into the impact of aspirin on the progression of esophageal cancer as discussed below.

Like that observed in other cancers associated with chronic inflammation, the progression of esophageal cancer is accompanied by an increase in the expression levels of COX-2.[34,43] This observation, as well as supportive evidence from pre-clinical studies, led to the clinical testing of specific inhibitors of COX-2 as well as the more non-specific NSAIDS, for their ability to inhibit esophageal cancer progression. The general consensus from the studies performed thus far is that specific COX-2 inhibitors like celecoxib are not effective at inhibiting progression of Barrett's esophagus. However, aspirin may be effective. The analyses of a number of studies that include randomized trials, case controls, and cohort studies indicate that long-term use of aspirin is associated with reduction in the incidences of esophageal cancer.[37] The major side effect that is of concern is an increased risk of bleeding which may lead to hemorrhagic strokes.

Epidemiological evidence also indicates that an inverse association exists between the use of statins and risk of developing esophageal adeno-carcinomas.[43] In the case-control and cohort studies examined, the effectiveness of statins was linked to both dose and duration. However, additional randomized clinical studies need to be performed to verify these results.

Flavonoids and esophageal cancer. Studies performed using cultured human esophageal carcinoma cell lines have shown that isorhamnetin,[44] luteolin,[45] myricetin,[46] baohuoside I,[47] and (−)-epigallocatechin-3-gallate (EGCG)[48] can inhibit cell growth and induce apoptosis. Analyses of rank order potencies revealed that quercetin and luteolin were the most, whereas myricetin was the least potent with respect to their abilities to induce growth arrest and apoptosis.[49] Further, EGCG was shown to inhibit the growth of cultured and explanted esophageal cells as well as in vitro cell invasion.[50] The underlying mechanisms appear to involve inhibition of ERK and COX-2 signaling. Using carcinogen-induced models of esophageal carcinogenesis, anthocyanins from black raspberries,[51] diosmin, hesperidin,[52] and EGCG[53] have been shown to inhibit either tumor formation rates, growth, and/or tumor multiplicity.

Epidemiological studies have reported some protective effects of flavonoid intake on the risk of developing esophageal cancer. For example, analyses of a northern Italian population revealed that of the different flavonoid classes, intake of only flavanones inversely correlated with esophageal cancer risk.[54] A study of black and white men residing in the United States found that consumption of anthocyanidin corresponded to a decreased risk of developing esophageal adenocarcinoma (by 53%) whereas consumption of isoflavonoids corresponded to a 57% decreased risk of developing esophageal squamous cell carcinoma but only in the white male population.[55] More recently, a large (521,448 subjects) European study examined the relationship between dietary ingestion of flavonoids and the risk of developing esophageal cancer.[56] Mean intake of total flavonoids in this population group was 434–446 mg/day with the highest consumption consisting of flavanols (approximately 350 mg/day) and the lowest of isoflavones (approximately 1.5 mg/day). Analyses of all human subjects revealed no association between intake of dietary flavonoids (either total or of any flavonoid class) and risk of esophageal cancer. However, analyses of only smokers revealed an inverse association between cancer risk and intake of total flavonoids, flavanols, and flavan-3-ol monomers.

Flavonoids and chemoprevention of esophageal cancer. Dietary, flavonoid-containing substances that have been tested for their impact on the progression of esophageal cancer include formulations containing berries. For example, a pilot study involving patients who were diagnosed with Barrett's esophagus, a condition of epithelial metaplasia that predisposes an individual to developing esophageal adenocarcinoma, was performed. The female and male subjects were administered lyophilized berry extract powder at 32 or 45 grams/day, respectively, for 26 weeks.[57] This was estimated to be comparable to a 5% dietary consumption or 1.5–2 cups of fresh berries. In five of the ten patients who ingested the berry powder, measures of oxidative damage markers were reduced, but the overall mean was similar to that of the control subjects and was not significantly different. A phase II clinical study also reported on the use of lyophilized strawberry powder as the putative chemopreventive agent.[58] Chemical analyses of the strawberry powder revealed that it contained a variety of chemopreventive agents of interest including vitamin C (488 milligram/100 gram), selenium (0.0131 milligram/100 gram), kaempferol (10.9 milligram/100 gram), and quercetin (5.12 milligram/100 gram). This powder was administered to patients who had been diagnosed with mild to moderate esophageal dysplasia. The 75 patients were randomly assigned to receive either 60 or 30 grams of powder, administered twice daily for six months. In the patients receiving the 30-gram dose, 5 of 36 (13.9%) were found to have a decreased grade of their precancerous lesions. A higher response rate (29 or 36 patients or 80.6%) was obtained in the patients receiving the 60-gram dose. Analyses of the esophageal mucosal biopsy samples revealed that the strawberry treatment resulted in decreases in the expression levels of proteins involve in mediating inflammation (i.e., iNOS, COX-2, and the p65 subunit of NFκB) and appeared to inhibit cell proliferation.

Chemoprevention of stomach cancer. Given the important role that *H. pylori* plays in the pathogenesis of stomach cancers, the majority of chemopreventive approaches have focused on inhibiting infection and inhibiting pathways associated with *H. pylori* infections, such as oxidative stress.[5] With this in mind, several clinical trials have been performed examining the effects of supplementation with antioxidants such as ascorbic acid, β-carotene, vitamin E, or selenium.[59] Collectively, these studies, performed in Columbia, Venezuela, and China which are all areas with high incidences of stomach cancer, have yielded inconsistent results. The agents capable of inhibiting COX-2 and inflammation, such as celecoxib and aspirin, have also been tested in several clinical trials using varying degrees of rigor. While some encouraging results indicating that low doses of aspirin may inhibit stomach cancer incidence, its use is not yet recommended for chemoprevention of gastric cancers.[37]

Flavonoids and stomach cancer. Studies performed using cultured human gastric cancer cells have shown that bufalin,[60] casticin,[61] daidzein,[62] galangin,[63] genistein,[62] luteolin,[62] nobiletin,[64] oroxylin A,[65] quercetin,[66] and a derivative of chrysin (5,7-dihydroxy-8-nitrochrysin)[67] can inhibit cell cycle progression and/or induce apoptosis. With respect to potential underlying mechanisms, casticin was shown to selectively enhance TRAIL/TNFSF10-induced cytotoxicity via enhanced ROS and increased expression of pro-apoptotic BAX.[61] In addition, the pro-apoptotic actions of bufalin appear to involve the PIK3CA/AKT pathway.[60] Finally, the mechanisms by which eupatilin inhibited in vitro cell invasion involved its inhibition of NFκB signaling.[68] From the epidemiological studies performed thus far, the evidence indicating a protective role of flavonoid consumption and the development of stomach cancer is quite limited.[69] Meta-analysis of 23 studies that included populations within Asia, Europe, and North America did not find an association between total flavonoid intake and risk of stomach cancer. However, when intake of specific flavonoid subclasses was analyzed, a significant inverse relationship was found between flavonal intake and stomach cancer risk.

Chemoprevention of hepatocellular carcinomas. Agents currently being considered for chemoprevention of hepatocellular carcinomas include vaccines and anti-viral therapies to prevent hepatitis B and/or hepatitis C infections.[39] In addition, the use of pharmacological agents such as statins and metformin are currently being tested for their ability to inhibit hepatocellular carcinoma progression. A number of epidemiological studies performed in patient populations residing in Taiwan, the United States, or Denmark who were considered to be at high risk for developing hepatocellular carcinomas have been performed. Collectively, these cohort and nested case control studies have observed that statin use is associated with a decrease in the incidence of hepatocellular carcinoma that was in some studies found to involve a dose-response relationship. However, a meta-analysis of 27 clinical trials investigating the effects of statin therapy indicated that there was a lack of evidence that statin therapy reduced the risk of any particular type of cancer, including that of hepatocellular carcinomas.[70] It has been proposed that statin users may also be taking drugs such as metformin which also has purported effects and this may mask the effects of statin therapy.[39] Epidemiological studies have also indicated that metformin use is associated with a decreased incidence of hepatocellular carcinomas.[39] The mechanisms that are thought to underlie the anti-cancer effects of metformin include its ability to block the mTOR pathway, inhibit the MYC proliferative pathway, inhibit the cell cycle regulator Cyclin D, and inhibit phosphorylation of RB thereby inhibiting cell proliferation.

Flavonoids and hepatocellular cancer. A number of studies have been performed using cultured hepatocellular carcinoma cells to demonstrate that flavonoids inhibit proliferation and/or induce apoptosis. Flavonoids thus far examined include apigenin,[71] baicalein,[72] breviscapine,[73] casticin,[74] chrysin,[75] (+)-cyanidan-3-ol,[76] galangin,[77] icaritin,[78] luteolin,[79] myricetin,[80] quercetin-3-o-glucoside,[81] and xanthohumol.[82] A study that examined eight isoflavones indicated that the ability of flavonoids to induce apoptosis may be structure-dependent.[83] Here, irilone was found to be the most potent with respect to inducing apoptosis in cultured human hepatocellular carcinoma cells (i.e., the Huh 7 cell line) whereas biochanin A and liquiritigenin were the least potent. In addition to inhibiting proliferation and inducing apoptosis, flavonoids such as baicalein[72] and galangin[84] have been shown to induce autophagy in cultured cells. The proposed underlying mechanisms by which these flavonoids are thought to inhibit proliferation and/or induce apoptosis include upregulation of TP53 signaling,[80] attenuation of STAT3 signaling,[79] induction of endoplasmic reticulum stress,[72,77] inactivation of FOXM1,[85] induction of the unfolded protein response,[75,86] induction of the death receptors, DR4 and DR5,[74] activation of MAPK8/JNK signaling,[78] induction of mitochondrial apoptosis,[87] and inhibition of the NFκB-mediated cell survival pathway.[82] Additional events monitored in cultured cells that are attributed to the chemoprevention activities of flavonoids include their protection against oxidative DNA damage induced by chemical carcinogens.[88] Flavonoids such as myricetin, quercetin, (+)-catechin and (−) epicatechin has been shown to be effective in this regard.

In addition to modifying cell fate processes, flavonoids such as baicalein have been shown to inhibit cell migration and invasion both in vitro and in vivo.[89] In vivo studies performed using explants of human hepatocellular carcinomas have reported that administration of baicalein[89] and luteolin[79] decreases the growth of the explanted tumors. Other in vivo models that have been used to demonstrate the chemopreventive activities of flavonoids include the use of N-nitrosodiethylamine/carbon tetrachloride[76] or diethylnitrosamine[90] to induce formation of the hepatocellular carcinomas. Using these models, co-administration of (+)-cyanidan-3-ol,[76] morin,[90] or 5,7 dimethyoxy flavones[91] was found to decrease the growth of the carcinogen-induced tumors and within the excised tumors, to increase the expression of genes involved in apoptosis.

Two recent epidemiological studies have reported on the relationship between the consumption of dietary flavonoids and the development of hepatocellular carcinomas.[92,93] Zomora-Ros *et al.* studied 477,206 subjects who were residents of ten western countries and found a borderline, but insignificant association, between the incidences of hepatocellular carcinomas and total flavonoid intake.[92] Over the 11.1-year period, 191 subjects developed hepatocellular carcinomas. The mean dietary intake of the subjects within this study was 438 mg/day. Intake of flavonols, the most important contributor to total flavonoid intake, was found to be inversely associated with hepatocellular carcinomas. In this study, 31% of the subjects who developed hepatocellular carcinomas were infected with either hepatitis B, hepatitis C, or both viruses. Lagiou *et al.* studied flavonoid intake and hepatocellular carcinomas in Greece.[93] In this three-year, case controlled study, the flavonoid intake of 339 subjects with liver cancer was compared to that of 360 control subjects. No relationship was found between either total flavonoid intake or with any class of flavonoids except with respect to flavones where an increased relationship was observed regardless of whether or not the subjects were infected with the hepatitis viruses. Total flavonoid intake in these subjects ranged from 146 to 358 mg/day. Given the accumulating evidence that flavonoids may exert anti-inflammatory properties, the extent to which flavonoids may protect against the development of hepatitis-associated hepatocellular carcinogenesis requires further study.

Chemoprevention of pancreatic cancer. Given the close relationship between chronic inflammation and the development of pancreatic cancer, chemopreventive approaches have focused largely on anti-inflammation strategies.[94,95] Pre-clinical and epidemiological studies support the idea that inhibition of COX-2 via the administration of either aspirin or NSAIDS could be effective at inhibiting or slowing the progression of pancreatic cancer. For example, a case control study was performed involving 904 patients with pancreatic ductal carcinomas and 1,224 matched controls.[96] No evidence of an association between use of acetaminophen or non-aspirin NSAIDS and risk of developing pancreatic cancer was observed. Use of aspirin, however, did appear to have a protective effect. Nonetheless, other studies have reported null associations between aspirin use and the development of pancreatic cancer. Metformin is also being considered for its potential as a chemopreventive agent for protecting against pancreatic cancer as its use is in general, supported by both epidemiological and pre-clinical studies.[94] Other potential chemopreventive agents include inhibitors of the renin-angiotensin system.

With respect to dietary agents, perhaps the best chemopreventive agent tested for its effectiveness at inhibiting pancreatic cancer is curcumin.[97,98] A phase II, non-randomized clinical trial involving 25 patients with confirmed adenocarcinoma of the pancreas was recently reported.[97] The patients received 8 grams of curcumin, administered orally. At the end of eight weeks, markers of inflammation (serum levels of IL6, IL8), after administration of oral curcumin were evaluated. The expression levels of markers of inflammation (NFκB, COX-2, and phosphorylated STAT3) were reduced in the blood mononuclear cells in the curcumin-treated patients. No significant adverse effects were reported. A second phase II clinical trial examined combined treatment of curcumin with gemcitabine.[98] The participating patients had been diagnosed with metastatic adenocarcinoma of the pancreas. The patients were orally administered 8 grams of curcumin each day. Eleven patients were able to complete the therapy which lasted until progression of the disease, one week to 12 months. The study was terminated early because of severe GI symptoms which included abdominal pain reported in five of the 11 patients. Of these patients, one exhibited a partial response, four were diagnosed with stable disease and six had tumor progression. These two studies using curcumin indicate that while high doses are well tolerated, when used in combination with chemotherapeutic agents, unexpected drug-supplement interactions can occur.

Flavonoids and pancreatic cancer. Studies performed in cultured pancreatic carcinoma cells have found that treatment with alpinetin,[99] apigenin,[100] fisetin,[101] myricetin,[102] or quercetin[103] can inhibit proliferation and/or induce apoptosis of these cells. The ability of apigenin,[100] hispidulin,[104] myricetin,[102] and quercetin[103] to inhibit the in vivo growth of explanted pancreatic cells has also been demonstrated. With respect to the mechanisms by which flavonoids may exert their anti-proliferative and pro-apoptotic effects, down regulation of the NFκB cell survival pathway appears to be involved as well as activation of JNK[105] (digitoflavone), inhibition of PIK3CA (myricetin),[102] and modulation of the DR3 death receptor (fisetin).[101] Additional mechanisms by which flavonoids may inhibit pancreatic carcinogenesis include their inhibition of angiogenesis. In this regard, in vivo administration of hispidulin has been shown to inhibit VEGF-induced angiogenesis within explanted tumor cells.[104]

Two recent epidemiological studies have examined the relationship between dietary intake of flavonoids and the incidence of pancreatic cancers. Bobe *et al.* (2008) studied a cohort of 50–69 males (27,111 subjects) who smoked at least five cigarettes/day and were residents of Finland.[106] During the course of the six-year study, 306 developed pancreatic cancer and total flavonoid intake varied from 6.45–22.88 mg/day. Total flavonoid intake was not significantly associated with the development of pancreatic cancer. A larger cohort (537,104 subjects) of men and women residing in the United States has also been reported.[107] In this ten-year study, 2,379 pancreatic cancers were identified and total flavonoid consumption ranged from 34.3 to 171.4 mg/day. No association between flavonoid consumption and pancreatic cancer incidence was found regardless of smoking status, BMI, or diabetes status.

Chemoprevention of colorectal cancer. Strategies used to prevent colorectal cancer include genetic counseling and dietary modifications designed to decrease the consumption of fats and red and processed meats while increasing the consumption of fiber and plant-based foods.[36,108,109] In addition, early detection methods have proven to be successful. Clinical trials designed to test the effectiveness of chemopreventive agents typically focus on individuals who are at high risk of developing colorectal cancer including those with Lynch syndrome (hereditary nonpolyposis colorectal cancer), polyposis, or IBD. The primary endpoint is typically colorectal adenomas. The majority of agents being tested are thought to act as anti-inflammatory agents; however, their potential activities on cancer stem cells, which may include inhibition of pathways such as the WNT/β-catenin pathways, should also be considered.

The most clinically proven, effective chemopreventive agents used to prevent or inhibit colorectal cancers are the NSAIDs which include aspirin, mesalazine, sulindac, and celecoxib. While mesalamine has been shown to exert chemopreventive effects with respect to colorectal cancer, its effectiveness in patients who do not have IBD has not yet been clearly demonstrated. Additional pharmacological agents that have been considered for their use in preventing colorectal cancers include statins, thiazolidinedione, and metformin. While the appropriate role of statins in chemoprevention of colorectal cancers is yet emerging, they may be most effective at delaying tumor progression and enhancing survival in patients diagnosed with adenomas. Thiazolidinediones are agonists of the peroxisome proliferator-activated receptor gamma and are often prescribed to control symptoms of diabetes. In addition to their ability to promote glucose and lipid homeostasis, thiazolidinediones inhibit the growth while enhancing apoptosis and terminal differentiation of colon tumor cells as well as inhibit inflammation. They may be most effective in diabetic patients who are at high risk for developing colorectal cancer. However, there is also evidence that these agents may exhibit pro-tumor activities and may increase the risk of developing heart failure, bone fractures, and perhaps bladder cancer. Metformin is often prescribed to control symptoms associated with diabetes as it is effective in inhibiting hepatic gluconeogenesis. In addition, it acts on the insulin/insulin-like growth factor/mTOR pathway. In addition to its effects on insulin/mTOR signaling, metformin may also selectively suppress colorectal stem cells via inhibition of NFκB and STAT3 signaling. While its chemopreventive effects are indicated in pre-clinical trials, convincing evidence has not yet been obtained from clinical trials.

In addition, dietary products that have been tested for their ability to prevent colorectal cancer include antioxidants (vitamin E, vitamin D, and β-carotene), micronutrients, and probiotics. Despite a number of clinical trials that have been performed testing the effectiveness of antioxidants, convincing evidence that their use may effectively inhibit colorectal cancers is lacking. In addition, in some cases, adverse effects including a possible increase in overall mortality have been reported. Clinical results testing the effectiveness of micronutrients such as folate, calcium, and selenium as well as probiotics have yielded inconsistent results.

Flavonoids and colorectal cancer. Flavonoids that have been shown to inhibit the in vitro proliferation and/or induce apoptosis of cultured colorectal tumor cells include apigenin,[110] baicalein,[111] fisetin,[112] hesperetin,[113] isorhamnetin,[114] luteolin,[115] methyl 3,5-dicaffeoyl quinate,[116] nobiletin,[117] oroxylin A,[118] silibinin,[119] and tangeretin.[117] The majority of these studies report that flavonoids both inhibit proliferation and induce apoptosis. However, nobiletin, tangeretin,[117] and silibinin[119] appear to only impact cell cycle control and induce cell cycle arrest (i.e., G1 arrest) without inducing apoptosis. With respect to mechanisms, the flavonoids studied have been shown to inhibit PIK3CA/AKT/mTOR signaling (isorhamnetin),[114] bind to ribosomal protein 9, a regulator of cyclin dependent kinase 1 expression (apigenin),[120] inhibit NFκB signaling (baicalein),[111] increase formation of reactive oxygen species and upregulate NRF2 signaling (oroxylin A),[121] inhibit EGFR and NFκB signaling (fisetin),[112] and enhance apoptosis via inhibition of uncoupling protein 2 (oroxylin A).[118]

In addition, apigenin has been shown to inhibit autophagy which is then thought to increase the cell's susceptibility to apoptotic-inducing agents.[110] Further, baicalein,[122] casticin,[123] and quercetin[124] have been shown to enhance apoptosis induced by TRAIL, a member of the TNF superfamily, via mechanisms that may include enhanced formation of reactive oxygen species.[123] Similarly, quercetin has been shown to enhance hypoxia-mediated apoptosis.[125] Further, silibinin has been shown to block the ability of the pro-inflammatory cytokine IL6 to induce apoptosis and cell invasion.[126]

Using cell explants, the administration of oroxylin A[121] and quercetin[125] has been shown to inhibit the in vivo growth of tumor cells. Additional animal models that have been used to examine the impact of flavonoids on colorectal tumorigenesis include carcinogen-induced colon tumors using chemicals such as azoxymethane (AOM) or 1,2-dimethylhydrazine (DMH). Using these animal models, isoliquiritigenin[127] and silymarin[128] have been found to reduce tumor formation initiated by AOM whereas treatment of apigenin and quercetin decreased the formation of preneoplastic lesions.[129,130] Interestingly, apigenin and naringenin, but not hesperidin and nobiletin, reduced formation of preneoplasic lesions.[131] However, Miyamoto *et al.* reported that chrysin, quercetin, and nobiletin were all capable of inhibiting the development of preneoplastic lesions in a genetically obese model.[132] Using the DMH model, hesperetin[133] and luteolin[134] decreased colon tumor formation and treatment with morin[135] decreased formation of preneoplastic lesions. Additional animal models used to study colorectal carcinogenesis are those developed to mimic the events associated with chronic inflammation that occurs within patients with IBD. Here, the mice are often administered dextran sodium sulfate to initiate colon inflammation, and are then exposed to a carcinogen, such as AOM or DMH, to initiate formation of the tumor cell. Using this approach, administration of isorhamnetin, but not myricetin, quercetin, or rutin, was shown to reduce tumor burden while decreasing markers of inflammation.[136] Similarly, baicalein,[111] nobiletin,[132] and silymarin[137] reduced tumorigenesis and silibinin reduced formation of preneoplastic lesions and expression levels of inflammatory markers.[138]

The relationship between human dietary intake of flavonoids and the risk of developing colorectal cancer has been examined in a number of studies. A case control study consisting of 264 cases of confirmed colo-rectal cancers and 408 controls was performed in a Scottish popula-tion.[139] The total dietary flavonoids consumed were flavonols (<19.3 to >40.41 mg/day), procyanidins (<21.3 to >49.81 ng/day), flavan-3-ols (<67.1 to >188.81 mg/day) and flavonones (<2.73 to >32.19 mg/day). No association between consumption of flavonols, procyanidins, or flavan-3-ol intake and risk of developing colorectal cancer was observed. However, there was a significant association between intake of quercetin and risk of developing colorectal cancer. A study of a population in the Netherlands composed of 58,279 men and 62,573 women used a 13.3-year follow-up where 1,444 male and 1,041 female subjects devel-oped colorectal cancer.[140] Here, total flavonol and flavone intake ranged from 1.4 to 105 mg/day and intake of catechin ranged from <36.2 to 287.3 mg/day. No association between total flavonol, flavone, or catechin intake and colorectal cancer was observed.

Flavonoids and colorectal cancer chemoprevention. Flavonoid-containing dietary products that have been studied for their potential as chemopreventive agents to treat colorectal cancer in human subjects include lyophilized berry extracts. A recently reported clinical phase I study used a lyophilized berry extract powder that contained 2,978 mg per 100 gram dry weight of total anthocyanins.[141] The patients participating in this study had been diagnosed with familial adenomatous polyposis and had at least five rectal polyps that were at least 2 mm in size. Twenty grams of berry powder was orally administered to patients three times per day for a total of 36 weeks. In addition, the patients received two rectal suppositories that were administered daily. The patient response to the berry powder included some adverse effects such as anal fissures, bloat-ing, diarrhea, flatulence, nausea, and rectal irritation. With respect to dis-ease progression, overall, a reduction in the number and burden of rectal adenomas was observed. However, not all of the patients responded. In the patients who did respond to the berry treatment, gene expression analyses of their biopsy samples obtained prior to and following the treatment regi-men indicated that the expression levels of genes involved in cell prolif-eration as well as overall global DNA methylation was reduced by the treatment with the lyophilized berry extracts.

Other flavonoid-containing interventions studied in patients diagnosed with colorectal cancer include a clinical trial composed of 87 patients, where flavonoids were administered in the form of two tablets containing 10 mg apigenin and 10 mg EGCG per tablet.[142] After two to five years, the combined rate of recurrence for neoplasia was 5 in 20 in the flavonoids-treated group as compared to 10 in 21 in the matched controls. While statistical significance was not achieved in this study (P = .20), as described by the authors, a trend of "favorable outcome" was observed. Additional clinical intervention studies that have examined the impact of flavonoids on events associated with colorectal cancer are those of Bobe[143] and Mentor-Marcel[144] that were discussed in Chapter 3.

References

1. Jemal A, Bray F, Center MM, Ferlay J, Ward E, Forman D. Global cancer statistics. *CA: A cancer journal for clinicians.* Mar–Apr 2011;61(2):69–90.
2. Boland CR. The molecular biology of gastrointestinal cancer: Implications for diagnosis and therapy. *Gastrointestinal endoscopy clinics of North America.* Jul 2008;18(3):401–413, vii.
3. Umar SD, Fleischer DE. Esophageal cancer: Epidemiology, pathogenesis and prevention. *Nature clinical practice. Gastroenterology & hepatology.* Sep 2008;5(9):517–526.
4. Gao YB, Chen ZL, Li JG, *et al.* Genetic landscape of esophageal squamous cell carcinoma. *Nature genetics.* Oct 2014;46(10):1097–1102.
5. Nagini S. Carcinoma of the stomach: A review of epidemiology, pathogenesis, molecular genetics and chemoprevention. *World journal of gastrointestinal oncology.* Jul 15, 2012;4(7):156–169.
6. Cervera P, Flejou JF. Changing pathology with changing drugs: Tumors of the gastrointestinal tract. *Pathobiology: Journal of immunopathology, molecular and cellular biology.* 2011;78(2):76–89.
7. Venook AP, Papandreou C, Furuse J, de Guevara LL. The incidence and epidemiology of hepatocellular carcinoma: A global and regional perspective. *The oncologist.* 2010;15 Suppl 4:5–13.
8. Arzumanyan A, Reis HM, Feitelson MA. Pathogenic mechanisms in HBV- and HCV-associated hepatocellular carcinoma. *Nature reviews. Cancer.* Feb 2013;13(2):123–135.

9. Kolodecik T, Shugrue C, Ashat M, Thrower EC. Risk factors for pancreatic cancer: Underlying mechanisms and potential targets. *Frontiers in physiology.* 2013;4:415.

10. Yeo TP, Lowenfels AB. Demographics and epidemiology of pancreatic cancer. *Cancer journal.* Nov–Dec 2012;18(6):477–484.

11. Pinho AV, Chantrill L, Rooman I. Chronic pancreatitis: A path to pancreatic cancer. *Cancer letters.* Apr 10, 2014;345(2):203–209.

12. Fleet JC. Animal models of gastrointestinal and liver diseases. New mouse models for studying dietary prevention of colorectal cancer. *American journal of physiology. Gastrointestinal and liver physiology.* Aug 1, 2014; 307(3):G249–259.

13. Linnekamp JF, Wang X, Medema JP, Vermeulen L. Colorectal Cancer Heterogeneity and Targeted Therapy: A Case for Molecular Disease Subtypes. *Cancer Res.* Jan 15, 2015;75(2):245–249.

14. Lynch HT, Snyder CL, Shaw TG, Heinen CD, Hitchins MP. Milestones of Lynch syndrome: 1895–2015. *Nature reviews. Cancer.* Mar 2015;15(3): 181–194.

15. Foersch S, Neurath MF. Colitis-associated neoplasia: Molecular basis and clinical translation. *Cellular and molecular life sciences: CMLS.* Sep 2014; 71(18):3523–3535.

16. Rogler G. Chronic ulcerative colitis and colorectal cancer. *Cancer letters.* Apr 10, 2014;345(2):235–241.

17. De Lerma Barbaro A, Perletti G, Bonapace IM, Monti E. Inflammatory cues acting on the adult intestinal stem cells and the early onset of cancer (review). *Int J Oncol.* Sep 2014;45(3):959–968.

18. Silveira MG, Lindor KD. Primary sclerosing cholangitis. *Canadian journal of gastroenterology = Journal canadien de gastroenterologie.* Aug 2008;22(8):689–698.

19. Marusawa H, Jenkins BJ. Inflammation and gastrointestinal cancer: An overview. *Cancer letters.* Apr 10, 2014;345(2):153–156.

20. Di Caro G, Marchesi F, Laghi L, Grizzi F. Immune cells: Plastic players along colorectal cancer progression. *Journal of cellular and molecular medicine.* Sep 2013;17(9):1088–1095.

21. Tye H, Jenkins BJ. Tying the knot between cytokine and toll-like receptor signaling in gastrointestinal tract cancers. *Cancer science.* Sep 2013; 104(9):1139–1145.

22. Kauppila JH, Selander KS. Toll-like receptors in esophageal cancer. *Frontiers in immunology.* 2014;5:200.
23. Castano-Rodriguez N, Kaakoush NO, Mitchell HM. Pattern-recognition receptors and gastric cancer. *Frontiers in immunology.* 2014;5:336.
24. Mahmoud AM, Zhu T, Parray A, *et al.* Differential effects of genistein on prostate cancer cells depend on mutational status of the androgen receptor. *PLoS One.* 2013;8(10):e78479.
25. Mohamed FE, Al-Jehani RM, Minogue SS, *et al.* Effect of toll-like receptor 7 and 9 targeted therapy to prevent the development of hepatocellular carcinoma. *Liver international: Official journal of the International Association for the Study of the Liver.* Mar 2015;35(3):1063–1076.
26. Zhang HF, Lai R. STAT3 in Cancer-Friend or Foe? *Cancers.* 2014;6(3): 1408–1440.
27. Mansell A, Jenkins BJ. Dangerous liaisons between interleukin-6 cytokine and toll-like receptor families: A potent combination in inflammation and cancer. *Cytokine & growth factor reviews.* Jun 2013;24(3):249–256.
28. Sheflin AM, Whitney AK, Weir TL. Cancer-promoting effects of microbial dysbiosis. *Current oncology reports.* Oct 2014;16(10):406.
29. Irrazabal T, Belcheva A, Girardin SE, Martin A, Philpott DJ. The multifaceted role of the intestinal microbiota in colon cancer. *Molecular cell.* Apr 24, 2014;54(2):309–320.
30. Abreu MT, Peek Jr. RM, Gastrointestinal malignancy and the microbiome. *Gastroenterology.* May 2014;146(6):1534–1546 e1533.
31. Keku TO, Dulal S, Deveaux A, Jovov B, Han X. The gastrointestinal microbiota and colorectal cancer. *American journal of physiology. Gastrointestinal and liver physiology.* Mar 1, 2015;308(5):G351-G363.
32. Louis P, Hold GL, Flint HJ. The gut microbiota, bacterial metabolites and colorectal cancer. *Nature reviews. Microbiology.* Oct 2014;12(10): 661–672.
33. Sears CL, Garrett WS. Microbes, microbiota, and colon cancer. *Cell host & microbe.* Mar 12, 2014;15(3):317–328.
34. Winberg H, Lindblad M, Lagergren J, Dahlstrand H. Risk factors and chemoprevention in Barrett's esophagus — an update. *Scandinavian journal of gastroenterology.* Apr 2012;47(4):397–406.
35. Chun KS, Kim EH, Lee S, Hahm KB. Chemoprevention of gastrointestinal cancer: The reality and the dream. *Gut and liver.* Mar 2013;7(2):137–149.

36. Stolfi C, De Simone V, Pallone F, Monteleone G. Mechanisms of action of non-steroidal anti-inflammatory drugs (NSAIDs) and mesalazine in the chemoprevention of colorectal cancer. *International journal of molecular sciences.* 2013;14(9):17972–17985.

37. Cuzick J, Thorat MA, Bosetti C, *et al.* Estimates of benefits and harms of prophylactic use of aspirin in the general population. *Annals of oncology: Official journal of the European Society for Medical Oncology/ESMO.* Jan 2015;26(1):47–57.

38. Patterson SL, Colbert Maresso K, Hawk E. Cancer chemoprevention: Successes and failures. *Clinical chemistry.* Jan 2013;59(1):94–101.

39. Singh S, Singh PP, Roberts LR, Sanchez W. Chemopreventive strategies in hepatocellular carcinoma. *Nature reviews. Gastroenterology & hepatology.* Jan 2014;11(1):45–54.

40. Obszynska JA, Atherfold PA, Nanji M, *et al.* Long-term proton pump induced hypergastrinaemia does induce lineage-specific restitution but not clonal expansion in benign Barrett's oesophagus in vivo. *Gut.* Feb 2010; 59(2):156–163.

41. Chang EY, Morris CD, Seltman AK, *et al.* The effect of antireflux surgery on esophageal carcinogenesis in patients with barrett esophagus: A systematic review. *Annals of surgery.* Jul 2007;246(1):11–21.

42. Das D, Chilton AP, Jankowski JA. Chemoprevention of oesophageal cancer and the AspECT trial. *Recent results in cancer research. Fortschritte der Krebsforschung. Progres dans les recherches sur le cancer.* 2009;181: 161–169.

43. Akiyama J, Alexandre L, Baruah A, *et al.* Strategy for prevention of cancers of the esophagus. *Ann N Y Acad Sci.* Sep 2014;1325:108–126.

44. Ma G, Yang C, Qu Y, Wei H, Zhang T, Zhang N. The flavonoid component isorhamnetin in vitro inhibits proliferation and induces apoptosis in Eca-109 cells. *Chem Biol Interact.* Apr 25, 2007;167(2):153–160.

45. Wang TT, Wang SK, Huang GL, Sun GJ. Luteolin induced-growth inhibition and apoptosis of human esophageal squamous carcinoma cell line Eca109 cells in vitro. *Asian Pacific journal of cancer prevention: APJCP.* 2012;13(11):5455–5461.

46. Zang W, Wang T, Wang Y, *et al.* Myricetin exerts anti-proliferative, anti-invasive, and pro-apoptotic effects on esophageal carcinoma EC9706 and KYSE30 cells via RSK2. *Tumour Biol.* Sep 6, 2014.

47. Wang L, Lu A, Liu X, *et al.* The flavonoid Baohuoside-I inhibits cell growth and downregulates survivin and cyclin D1 expression in esophageal carcinoma via beta-catenin-dependent signaling. *Oncol Rep.* Nov 2011; 26(5):1149–1156.

48. Hou Z, Sang S, You H, *et al.* Mechanism of action of (−)-epigallocatechin-3-gallate: Auto-oxidation-dependent inactivation of epidermal growth factor receptor and direct effects on growth inhibition in human esophageal cancer KYSE 150 cells. *Cancer Res.* Sep 1, 2005;65(17):8049–8056.

49. Zhang Q, Zhao XH, Wang ZJ. Flavones and flavonols exert cytotoxic effects on a human oesophageal adenocarcinoma cell line (OE33) by causing G2/M arrest and inducing apoptosis. *Food Chem Toxicol.* Jun 2008;46(6):2042–2053.

50. Ye F, Zhang GH, Guan BX, Xu XC. Suppression of esophageal cancer cell growth using curcumin, (−)-epigallocatechin-3-gallate and lovastatin. *World journal of gastroenterology: WJG.* Jan 14, 2012;18(2):126–135.

51. Wang LS, Hecht SS, Carmella SG, *et al.* Anthocyanins in black raspberries prevent esophageal tumors in rats. *Cancer Prev Res (Phila).* Jan 2009; 2(1):84–93.

52. Tanaka T, Makita H, Kawabata K, *et al.* Modulation of N-methyl-N-amylnitrosamine-induced rat oesophageal tumourigenesis by dietary feeding of diosmin and hesperidin, both alone and in combination. *Carcinogenesis.* Apr 1997;18(4):761–769.

53. Morse MA, Kresty LA, Steele VE, *et al.* Effects of theaflavins on N-nitrosomethylbenzylamine-induced esophageal tumorigenesis. *Nutr Cancer.* 1997;29(1):7–12.

54. Rossi M, Garavello W, Talamini R, *et al.* Flavonoids and risk of squamous cell esophageal cancer. *Int J Cancer.* Apr 1, 2007;120(7):1560–1564.

55. Bobe G, Peterson JJ, Gridley G, Hyer M, Dwyer JT, Brown LM. Flavonoid consumption and esophageal cancer among black and white men in the United States. *Int J Cancer.* Sep 1, 2009;125(5):1147–1154.

56. Vermeulen E, Zamora-Ros R, Duell EJ, *et al.* Dietary flavonoid intake and esophageal cancer risk in the European prospective investigation into cancer and nutrition cohort. *American journal of epidemiology.* Aug 15, 2013; 178(4):570–581.

57. Kresty LA, Frankel WL, Hammond CD, *et al.* Transitioning from preclinical to clinical chemopreventive assessments of lyophilized black raspberries: Interim results show berries modulate markers of oxidative stress in Barrett's esophagus patients. *Nutr Cancer.* 2006;54(1):148–156.

58. Chen T, Yan F, Qian J, *et al*. Randomized phase II trial of lyophilized straw-berries in patients with dysplastic precancerous lesions of the esophagus. *Cancer Prev Res (Phila)*. Jan 2012;5(1):41–50.

59. Pasechnikov V, Chukov S, Fedorov E, Kikuste I, Leja M. Gastric cancer: Prevention, screening and early diagnosis. *World journal of gastroenterology: WJG*. Oct 14, 2014;20(38):13842–13862.

60. Li D, Qu X, Hou K, *et al*. PI3K/Akt is involved in bufalin-induced apoptosis in gastric cancer cells. *Anti-cancer drugs*. Jan 2009;20(1):59–64.

61. Zhou Y, Tian L, Long L, Quan M, Liu F, Cao J. Casticin potentiates TRAIL-induced apoptosis of gastric cancer cells through endoplasmic reticulum stress. *PLoS One*. 2013;8(3):e58855.

62. Matsukawa Y, Marui N, Sakai T, *et al*. Genistein arrests cell cycle progression at G2-M. *Cancer Res*. Mar 15, 1993;53(6):1328–1331.

63. Kim DA, Jeon YK, Nam MJ. Galangin induces apoptosis in gastric cancer cells via regulation of ubiquitin carboxy-terminal hydrolase isozyme L1 and glutathione S-transferase P. *Food Chem Toxicol*. Mar 2012;50(3–4):684–688.

64. Moon JY, Cho M, Ahn KS, Cho SK. Nobiletin induces apoptosis and poten-tiates the effects of the anticancer drug 5-fluorouracil in p53-mutated SNU-16 human gastric cancer cells. *Nutr Cancer*. 2013;65(2):286–295.

65. Yang Y, Hu Y, Gu HY, *et al*. Oroxylin A induces G2/M phase cell-cycle arrest via inhibiting Cdk7-mediated expression of Cdc2/p34 in human gas-tric carcinoma BGC-823 cells. *The Journal of pharmacy and pharmacology*. Nov 2008;60(11):1459–1463.

66. Yoshida M, Sakai T, Hosokawa N, *et al*. The effect of quercetin on cell cycle progression and growth of human gastric cancer cells. *FEBS letters*. Jan 15 1990;260(1):10–13.

67. Ai XH, Zheng X, Tang XQ, *et al*. Induction of apoptosis of human gastric carcinoma SGC-7901 cell line by 5, 7-dihydroxy-8-nitrochrysin in vitro. *World journal of gastroenterology: WJG*. Jul 28, 2007;13(28):3824–3828.

68. Park BB, Yoon J, Kim E, *et al*. Inhibitory effects of eupatilin on tumor invasion of human gastric cancer MKN-1 cells. *Tumour Biol*. Apr 2013;34(2):875–885.

69. Woo HD, Kim J. Dietary flavonoid intake and risk of stomach and colorectal cancer. *World journal of gastroenterology: WJG*. Feb 21, 2013;19(7):1011–1019.

70. Cholesterol Treatment Trialists C, Emberson JR, Kearney PM, *et al*. Lack of effect of lowering LDL cholesterol on cancer: Meta-analysis of individual data from 175,000 people in 27 randomised trials of statin therapy. *PLoS One*. 2012;7(1):e29849.

71. Choi SI, Jeong CS, Cho SY, Lee YS. Mechanism of apoptosis induced by apigenin in HepG2 human hepatoma cells: Involvement of reactive oxygen species generated by NADPH oxidase. *Archives of pharmacal research.* Oct 2007;30(10):1328–1335.

72. Wang Z, Jiang C, Chen W, *et al.* Baicalein induces apoptosis and autophagy via endoplasmic reticulum stress in hepatocellular carcinoma cells. *BioMed research international.* 2014;2014:732516.

73. Wu Y, Fan Q, Lu N, *et al.* Breviscapine-induced apoptosis of human hepatocellular carcinoma cell line HepG2 was involved in its antitumor activity. *Phytother Res.* Aug 2010;24(8):1188–1194.

74. Yang J, Yang Y, Tian L, Sheng XF, Liu F, Cao JG. Casticin-induced apoptosis involves death receptor 5 upregulation in hepatocellular carcinoma cells. *World journal of gastroenterology: WJG.* Oct 14, 2011;17(38):4298–4307.

75. Sun X, Huo X, Luo T, Li M, Yin Y, Jiang Y. The anticancer flavonoid chrysin induces the unfolded protein response in hepatoma cells. *Journal of cellular and molecular medicine.* Nov 2011;15(11):2389–2398.

76. Monga J, Pandit S, Chauhan RS, Chauhan CS, Chauhan SS, Sharma M. Growth inhibition and apoptosis induction by (+)-Cyanidan-3-ol in hepatocellular carcinoma. *PLoS One.* 2013;8(7):e68710.

77. Su L, Chen X, Wu J, *et al.* Galangin inhibits proliferation of hepatocellular carcinoma cells by inducing endoplasmic reticulum stress. *Food Chem Toxicol.* Dec 2013;62:810–816.

78. He J, Wang Y, Duan F, Jiang H, Chen MF, Tang SY. Icaritin induces apoptosis of HepG2 cells via the JNK1 signaling pathway independent of the estrogen receptor. *Planta medica.* Nov 2010;76(16):1834–1839.

79. Selvendiran K, Koga H, Ueno T, *et al.* Luteolin promotes degradation in signal transducer and activator of transcription 3 in human hepatoma cells: An implication for the antitumor potential of flavonoids. *Cancer Res.* May 1, 2006;66(9):4826–4834.

80. Zhang XH, Zou ZQ, Xu CW, Shen YZ, Li D. Myricetin induces G2/M phase arrest in HepG2 cells by inhibiting the activity of the cyclin B/Cdc2 complex. *Mol Med Rep.* Mar–Apr 2011;4(2):273–277.

81. Sudan S, Rupasinghe HP. Quercetin-3-O-glucoside induces human DNA topoisomerase II inhibition, cell cycle arrest and apoptosis in hepatocellular carcinoma cells. *Anticancer research.* Apr 2014;34(4):1691–1699.

82. Dorn C, Kraus B, Motyl M, *et al.* Xanthohumol, a chalcon derived from hops, inhibits hepatic inflammation and fibrosis. *Mol Nutr Food Res.* Jul 2010;54 Suppl 2:S205–213.

83. Mansoor TA, Ramalho RM, Luo X, Ramalhete C, Rodrigues CM, Ferreira MJ. Isoflavones as apoptosis inducers in human hepatoma HuH-7 cells. *Phytother Res.* Dec 2011;25(12):1819–1824.

84. Wen M, Wu J, Luo H, Zhang H. Galangin induces autophagy through upregulation of p53 in HepG2 cells. *Pharmacology.* 2012;89(5–6):247–255.

85. He L, Yang X, Cao X, Liu F, Quan M, Cao J. Casticin induces growth suppression and cell cycle arrest through activation of FOXO3a in hepatocellular carcinoma. *Oncol Rep.* Jan 2013;29(1):103–108.

86. Xu M, Lu N, Sun Z, *et al.* Activation of the unfolded protein response contributed to the selective cytotoxicity of oroxylin A in human hepatocellular carcinoma HepG2 cells. *Toxicol Lett.* Jul 20, 2012;212(2):113–125.

87. Liu W, Mu R, Nie FF, *et al.* MAC-related mitochondrial pathway in oroxylin-A-induced apoptosis in human hepatocellular carcinoma HepG2 cells. *Cancer letters.* Nov 1, 2009;284(2):198–207.

88. Delgado ME, Haza AI, Arranz N, Garcia A, Morales P. Dietary polyphenols protect against N-nitrosamines and benzo(a)pyrene-induced DNA damage (strand breaks and oxidized purines/pyrimidines) in HepG2 human hepatoma cells. *Eur J Nutr.* Dec 2008;47(8):479–490.

89. Chiu YW, Lin TH, Huang WS, *et al.* Baicalein inhibits the migration and invasive properties of human hepatoma cells. *Toxicol Appl Pharmacol.* Sep 15, 2011;255(3):316–326.

90. Sivaramakrishnan V, Devaraj SN. Morin fosters apoptosis in experimental hepatocellular carcinogenesis model. *Chem Biol Interact.* Jan 27, 2010; 183(2):284–292.

91. Khan MS, Halagowder D, Devaraj SN. Methylated chrysin, a dimethoxy flavone, partially suppresses the development of liver preneoplastic lesions induced by N-nitrosodiethylamine in rats. *Food Chem Toxicol.* Jan 2011;49(1):173–178.

92. Zamora-Ros R, Fedirko V, Trichopoulou A, *et al.* Dietary flavonoid, lignan and antioxidant capacity and risk of hepatocellular carcinoma in the European prospective investigation into cancer and nutrition study. *Int J Cancer.* Nov 15, 2013;133(10):2429–2443.

93. Lagiou P, Rossi M, Lagiou A, Tzonou A, La Vecchia C, Trichopoulos D. Flavonoid intake and liver cancer: A case-control study in Greece. *Cancer causes & control: CCC.* Oct 2008;19(8):813–818.

94. Fendrich V. Chemoprevention of pancreatic cancer-one step closer. *Langenbeck's archives of surgery / Deutsche Gesellschaft fur Chirurgie.* Apr 2012;397(4):495–505.

95. Yue W, Yang CS, DiPaola RS, Tan XL. Repurposing of metformin and aspirin by targeting AMPK-mTOR and inflammation for pancreatic cancer prevention and treatment. *Cancer Prev Res (Phila).* Apr 2014;7(4):388–397.

96. Tan XL, Reid Lombardo KM, Bamlet WR, *et al.* Aspirin, nonsteroidal anti-inflammatory drugs, acetaminophen, and pancreatic cancer risk: A clinic-based case-control study. *Cancer Prev Res (Phila).* Nov 2011; 4(11):1835–1841.

97. Dhillon N, Aggarwal BB, Newman RA, *et al.* Phase II trial of curcumin in patients with advanced pancreatic cancer. *Clinical cancer research: An official journal of the American Association for Cancer Research.* Jul 15, 2008;14(14):4491–4499.

98. Epelbaum R, Schaffer M, Vizel B, Badmaev V, Bar-Sela G. Curcumin and gemcitabine in patients with advanced pancreatic cancer. *Nutr Cancer.* 2010;62(8):1137–1141.

99. Du J, Tang B, Wang J, *et al.* Antiproliferative effect of alpinetin in BxPC-3 pancreatic cancer cells. *International journal of molecular medicine.* Apr 2012;29(4):607–612.

100. King JC, Lu QY, Li G, *et al.* Evidence for activation of mutated p53 by apigenin in human pancreatic cancer. *Biochim Biophys Acta.* Feb 2012; 1823(2):593–604.

101. Murtaza I, Adhami VM, Hafeez BB, Saleem M, Mukhtar H. Fisetin, a natural flavonoid, targets chemoresistant human pancreatic cancer AsPC-1 cells through DR3-mediated inhibition of NF-kappaB. *Int J Cancer.* Nov 15, 2009;125(10):2465–2473.

102. Phillips PA, Sangwan V, Borja-Cacho D, Dudeja V, Vickers SM, Saluja AK. Myricetin induces pancreatic cancer cell death via the induction of apoptosis and inhibition of the phosphatidylinositol 3-kinase (PI3K) signaling pathway. *Cancer letters.* Sep 28, 2011;308(2):181–188.

103. Angst E, Park JL, Moro A, *et al.* The flavonoid quercetin inhibits pancreatic cancer growth in vitro and in vivo. *Pancreas.* Mar 2013;42(2):223–229.

104. He L, Wu Y, Lin L, *et al*. Hispidulin, a small flavonoid molecule, suppresses the angiogenesis and growth of human pancreatic cancer by targeting vascular endothelial growth factor receptor 2-mediated PI3K/Akt/mTOR signaling pathway. *Cancer science*. Jan 2011;102(1):219–225.

105. Cai X, Lu W, Yang Y, *et al*. Digitoflavone inhibits IkappaBalpha kinase and enhances apoptosis induced by TNFalpha through downregulation of expression of nuclear factor kappaB-regulated gene products in human pancreatic cancer cells. *PLoS One*. 2013;8(10):e77126.

106. Bobe G, Weinstein SJ, Albanes D, *et al*. Flavonoid intake and risk of pancreatic cancer in male smokers (Finland). *Cancer epidemiology, biomarkers & prevention: A publication of the American Association for Cancer Research, cosponsored by the American Society of Preventive Oncology*. Mar 2008;17(3):553–562.

107. Arem H, Bobe G, Sampson J, *et al*. Flavonoid intake and risk of pancreatic cancer in the National Institutes of Health-AARP Diet and Health Study Cohort. *British journal of cancer*. Mar 19, 2013;108(5):1168–1172.

108. Kim TI. Chemopreventive drugs: Mechanisms via inhibition of cancer stem cells in colorectal cancer. *World journal of gastroenterology: WJG*. Apr 14, 2014;20(14):3835–3846.

109. Nolfo F, Rametta S, Marventano S, *et al*. Pharmacological and dietary prevention for colorectal cancer. *BMC surgery*. 2013;13 Suppl 2:S16.

110. Lee Y, Sung B, Kang YJ, *et al*. Apigenin-induced apoptosis is enhanced by inhibition of autophagy formation in HCT116 human colon cancer cells. *Int J Oncol*. May 2014;44(5):1599–1606.

111. Kim DH, Hossain MA, Kang YJ, *et al*. Baicalein, an active component of Scutellaria baicalensis Georgi, induces apoptosis in human colon cancer cells and prevents AOM/DSS-induced colon cancer in mice. *Int J Oncol*. Nov 2013;43(5):1652–1658.

112. Suh Y, Afaq F, Johnson JJ, Mukhtar H. A plant flavonoid fisetin induces apoptosis in colon cancer cells by inhibition of COX2 and Wnt/EGFR/NF-kappaB-signaling pathways. *Carcinogenesis*. Feb 2009;30(2):300–307.

113. Sivagami G, Vinothkumar R, Bernini R, *et al*. Role of hesperetin (a natural flavonoid) and its analogue on apoptosis in HT-29 human colon adenocarcinoma cell line — a comparative study. *Food Chem Toxicol*. Mar 2012;50(3–4):660–671.

114. Li C, Yang X, Chen C, Cai S, Hu J. Isorhamnetin suppresses colon cancer cell growth through the PI3KAktmTOR pathway. *Mol Med Rep.* Mar 2014;9(3):935–940.

115. Ramos AA, Pereira-Wilson C, Collins AR. Protective effects of ursolic acid and luteolin against oxidative DNA damage include enhancement of DNA repair in Caco-2 cells. *Mutation research.* Oct 13, 2010;692(1–2):6–11.

116. Hu W, Shen T, Wang MH. Cell cycle arrest and apoptosis induced by methyl 3,5-dicaffeoyl quinate in human colon cancer cells: Involvement of the PI3K/Akt and MAP kinase pathways. *Chem Biol Interact.* Oct 15, 2011; 194(1):48–57.

117. Morley KL, Ferguson PJ, Koropatnick J. Tangeretin and nobiletin induce G1 cell cycle arrest but not apoptosis in human breast and colon cancer cells. *Cancer letters.* Jun 18, 2007;251(1):168–178.

118. Qiao C, Wei L, Dai Q, *et al.* UCP2-related mitochondrial pathway participates in oroxylin A-induced apoptosis in human colon cancer cells. *J Cell Physiol.* Sep 24, 2014.

119. Hogan FS, Krishnegowda NK, Mikhailova M, Kahlenberg MS. Flavonoid, silibinin, inhibits proliferation and promotes cell-cycle arrest of human colon cancer. *The Journal of surgical research.* Nov 2007;143(1):58–65.

120. Iizumi Y, Oishi M, Taniguchi T, Goi W, Sowa Y, Sakai T. The flavonoid apigenin downregulates CDK1 by directly targeting ribosomal protein S9. *PLoS One.* 2013;8(8):e73219.

121. Hu R, Chen N, Yao J, *et al.* The role of Nrf2 and apoptotic signaling pathways in oroxylin A-mediated responses in HCT-116 colorectal adenocarcinoma cells and xenograft tumors. *Anti-cancer drugs.* Jul 2012;23(6):651–658.

122. Taniguchi H, Yoshida T, Horinaka M, *et al.* Baicalein overcomes tumor necrosis factor-related apoptosis-inducing ligand resistance via two different cell-specific pathways in cancer cells but not in normal cells. *Cancer Res.* Nov 1, 2008;68(21):8918–8927.

123. Tang SY, Zhong MZ, Yuan GJ, *et al.* Casticin, a flavonoid, potentiates TRAIL-induced apoptosis through modulation of anti-apoptotic proteins and death receptor 5 in colon cancer cells. *Oncol Rep.* Feb 2013;29(2):474–480.

124. Psahoulia FH, Drosopoulos KG, Doubravska L, Andera L, Pintzas A. Quercetin enhances TRAIL-mediated apoptosis in colon cancer cells by inducing the accumulation of death receptors in lipid rafts. *Molecular cancer therapeutics.* Sep 2007;6(9):2591–2599.

125. Kim HS, Wannatung T, Lee S, *et al.* Quercetin enhances hypoxia-mediated apoptosis via direct inhibition of AMPK activity in HCT116 colon cancer. *Apoptosis.* Sep 2012;17(9):938–949.

126. Lin CM, Chen YH, Ma HP, *et al.* Silibinin inhibits the invasion of IL-6-stimulated colon cancer cells via selective JNK/AP-1/MMP-2 modulation in vitro. *J Agric Food Chem.* Dec 26, 2012;60(51):12451–12457.

127. Zhao H, Zhang X, Chen X, *et al.* Isoliquiritigenin, a flavonoid from licorice, blocks M2 macrophage polarization in colitis-associated tumorigenesis through downregulating PGE2 and IL-6. *Toxicol Appl Pharmacol.* Sep 15, 2014;279(3):311–321.

128. Kohno H, Tanaka T, Kawabata K, *et al.* Silymarin, a naturally occurring polyphenolic antioxidant flavonoid, inhibits azoxymethane-induced colon carcinogenesis in male F344 rats. *Int J Cancer.* Oct 10, 2002;101(5): 461–468.

129. Au A, Li B, Wang W, Roy H, Koehler K, Birt D. Effect of dietary apigenin on colonic ornithine decarboxylase activity, aberrant crypt foci formation, and tumorigenesis in different experimental models. *Nutr Cancer.* 2006;54(2):243–251.

130. Warren CA, Paulhill KJ, Davidson LA, *et al.* Quercetin may suppress rat aberrant crypt foci formation by suppressing inflammatory mediators that influence proliferation and apoptosis. *The Journal of nutrition.* Jan 2009;139(1):101–105.

131. Leonardi T, Vanamala J, Taddeo SS, *et al.* Apigenin and naringenin suppress colon carcinogenesis through the aberrant crypt stage in azoxymethane-treated rats. *Experimental biology and medicine.* Jun 2010;235(6):710–717.

132. Miyamoto S, Yasui Y, Ohigashi H, Tanaka T, Murakami A. Dietary flavonoids suppress azoxymethane-induced colonic preneoplastic lesions in male C57BL/KsJ-db/db mice. *Chem Biol Interact.* Jan 27, 2010;183(2):276–283.

133. Aranganathan S, Nalini N. Efficacy of the potential chemopreventive agent, hesperetin (citrus flavanone), on 1,2-dimethylhydrazine induced colon carcinogenesis. *Food Chem Toxicol.* Oct 2009;47(10):2594–2600.

134. Manju V, Nalini N. Protective role of luteolin in 1,2-dimethylhydrazine induced experimental colon carcinogenesis. *Cell biochemistry and function.* Mar–Apr 2007;25(2):189–194.

135. Karthik Kumar V, Vennila S, Nalini N. Inhibitory effect of morin on DMH-induced biochemical changes and aberrant crypt foci formation in

experimental colon carcinogenesis. *Environmental toxicology and pharmacology.* Jan 2010;29(1):50–57.

136. Saud SM, Young MR, Jones-Hall YL, *et al.* Chemopreventive activity of plant flavonoid isorhamnetin in colorectal cancer is mediated by oncogenic Src and beta-catenin. *Cancer Res.* Sep 1, 2013;73(17):5473–5484.

137. Toyoda-Hokaiwado N, Yasui Y, Muramatsu M, *et al.* Chemopreventive effects of silymarin against 1,2-dimethylhydrazine plus dextran sodium sulfate-induced inflammation-associated carcinogenicity and genotoxicity in the colon of gpt delta rats. *Carcinogenesis.* Oct 2011;32(10):1512–1517.

138. Kauntz H, Bousserouel S, Gosse F, Marescaux J, Raul F. Silibinin, a natural flavonoid, modulates the early expression of chemoprevention biomarkers in a preclinical model of colon carcinogenesis. *Int J Oncol.* Sep 2012; 41(3):849–854.

139. Kyle JA, Sharp L, Little J, Duthie GG, McNeill G. Dietary flavonoid intake and colorectal cancer: A case-control study. *The British journal of nutrition.* Feb 2010;103(3):429–436.

140. Simons CC, Hughes LA, Arts IC, Goldbohm RA, van den Brandt PA, Weijenberg MP. Dietary flavonol, flavone and catechin intake and risk of colorectal cancer in the Netherlands Cohort Study. *Int J Cancer.* Dec 15, 2009;125(12):2945–2952.

141. Wang LS, Burke CA, Hasson H, *et al.* A phase Ib study of the effects of black raspberries on rectal polyps in patients with familial adenomatous polyposis. *Cancer Prev Res (Phila).* Jul 2014;7(7):666–674.

142. Hoensch H, Groh B, Edler L, Kirch W. Prospective cohort comparison of flavonoid treatment in patients with resected colorectal cancer to prevent recurrence. *World journal of gastroenterology: WJG.* Apr 14, 2008; 14(14):2187–2193.

143. Bobe G, Murphy G, Albert PS, *et al.* Serum cytokine concentrations, flavonol intake and colorectal adenoma recurrence in the Polyp Prevention Trial. *British journal of cancer.* Oct 26, 2010;103(9):1453–1461.

144. Mentor-Marcel RA, Bobe G, Sardo C, *et al.* Plasma cytokines as potential response indicators to dietary freeze-dried black raspberries in colorectal cancer patients. *Nutr Cancer.* Aug 2012;64(6):820–825.

Flavonoids and Steroid-responsive Cancers*   **5**

Steroid-responsive cancers. Steroid-responsive cancers, also referred to as endocrine-related cancers, have historically included cancers of the breast, endometrium, ovary, prostate, and testis, as well as the thyroid and pituitary.[1] Given the paucity of evidence linking flavonoids to thyroid or pituitary cancers, these two types of cancers will not be subject to further discussion. As classically defined, the development and progression of these cancers is dependent on the absence or presence of steroid hormones. Hence, increased lifelong exposure to a particular sex steroid is a

---

** **Abbreviations:** AKT, v-akt murine thymoma viral oncogene homolog 1, also known as protein kinase B; AR, androgen receptor; ARID1A, AT-rich interacting domain-containing protein 1A; BAX, BCL2-associated X protein; BRAF, proto-oncogene B-Raf; BRCA1, breast cancer 1, early onset; BRCA2, breast cancer 2, early onset; CDKN2A, cyclin-dependent kinase inhibitor 2A, also known as p16; DMBA, 7,12 dimethylbenz(a)anthracene; EGCG, (-)-epigallocatechin-3-gallate; EF-1α, elongation factor 1 alpha; EGFR, epidermal growth factor receptor; ERα, estrogen receptor α; ERβ, estrogen receptor β;ERBB2, erb-b2 receptor tyrosine kinase 2; ERK, extracellular signal regulated kinase; FAK, focal adhesion kinase; FAS, Fas cell surface death receptor; HER2, human epidermal receptor growth factor receptor 2; hnRNPA1, heterogeneous nuclear ribonucleoprotein A1; IGF-1, insulin-like growth factor 1; KRAS, Kirsten rat sarcoma viral oncogene homolog; MAPK, mitogen activated protein kinase; MET, met proto-oncogene; MMP2/9, matrix metallopeptidase 2/9; mTOR, mechanistic target of rapamycin; MYC, v-myc avian myelocytomatosis viral oncogene homolog; NFκB, nuclear factor of kappa light polypeptide gene enhancer in B-cells 1; NSAID, non-steroidal anti-inflammatory drug; PIK3CA, phosphatidylinositol-4,5-bisphosphate 3-kinase catalytic subunit alpha; PIN, intraepithelial neoplasia; PR, progesterone receptor; PSA, prostate-specific antigen; PTEN, phosphatase and tensin homolog; RANKL, receptor activator of nuclear factor kappa-B; STK11/LKB1, serine/threonine kinase 11; TNFα, tumor necrosis factor alpha; TP53, tumor protein 53, TRAIL/TNFSF10, tumor necrosis factor (ligand) superfamily, member 10; TRAMP, transgenic adenocarcinoma of the mouse prostate; VEGF, vascular endothelial growth factor; WNT, wingless-type MMTV integration site family.*

major risk factor for steroid-responsive cancers and its absence typically forms the basis of very effective chemotherapeutics. For example, high, uninterrupted lifelong exposure to estrogen that arises from early onset of menarche, delayed menopause, and the absence of childbearing and lactation significantly increases a woman's risk of developing breast cancer. Accordingly, amongst the most effective breast cancer treatments are those that block the actions of the estrogen receptor. The major steroid hormone receptors under consideration are the estrogen, progesterone, and androgen receptors which are members of the nuclear steroid receptor superfamily.

Agonist activation of steroid hormone receptors results in their binding to specific DNA sequences typically found upstream of the promoters of their target genes. As a consequence, increased as well as decreased expression of a panel of steroid-responsive genes ensues. The extent to which these ligand-bound steroid receptors regulate activities of their respective target genes is determined by their interactions with coactivator and corepressor proteins that are often expressed at varying levels within specific tissues and within particular cell types. In addition to this classical "genomic" mechanism, the steroid receptors participate in a non-genomic mechanism that involves membrane-bound forms of steroid receptors and upregulation of kinase signaling cascades. While steroid hormones are well characterized for their key roles in breast, ovarian, uterine, and prostate cancers, it is becoming increasingly clear that these hormones can significantly impact the risk and progression of most cancers and may in fact, underlie the cancer risk posed by other disease states such as obesity. Thus, the ability of certain flavonoids to interact with and block the actions of hormone receptors, in particular, the estrogen, progesterone, and androgen receptors is of increasingly high interest.

Estrogen receptors and cancer. The major circulating estrogen is 17β estradiol that is produced primarily in the ovaries in the female. In the male, significant sources of estradiol as well as high expression levels of estrogen receptors are found throughout the male reproductive tract, in particular, the Leydig and germ cells.[2] Two estrogen receptor (ER) genes exist that encode estrogen receptor α (ERα) and estrogen receptor β (ERβ) which differ structurally, mechanistically, and with respect to ligand binding preferences.[1] ERα is thought to drive the pro-proliferative response, particularly in responsive cells within the breast, ovary, and endometrium whereas ERβ exerts predominantly anti-proliferative activities. In addition, ERβ is thought to be involved in the migratory activities of malignant tumor cells. High expression of ERα is found in the mammary gland, uterus, ovary, bone, testes, liver, and adipose tissue.[3] The expression of ERβ appears to be more restricted and is found at relatively high levels within the prostate, ovary, colon, adipose tissue, and immune system. Since approximately 60% of breast and ovarian cancers express ERα, a major driver of tumor progression in these tissues, treatments of these cancers typically involve **inhibitors** of ERα.[4] However, ongoing research efforts indicate that **activators** of ERβ may be effective for the treatment of certain breast and prostate cancers.

Progesterone receptor and breast cancer. As compared to that of the ER, the role of the progesterone receptor (PR) in the development of breast cancer and as a breast cancer drug target is under-appreciated.[5,6] Because the PR is a target gene of the ER, the two receptors are often coordinately expressed. Although they both regulate cell proliferation within the luminal cells of the breast tissue, they appear to play overlapping but distinct roles. Like the ER, the PR participates in not only genomic actions, but also extensive crosstalk with growth factor receptor pathways (i.e., ERBB2/HER2) that contribute to MAPK-dependent cell cycle progression. Important mediators of genomic PR signaling are amphiregulin, WNT4 and RANKL (receptor activator of NFκB, a TNFα family member). In animal models, the impact of PR on tumor progression can be demonstrated by administering a PR agonist, such as medroxy-progesterone acetate, to promote the formation of carcinogen-initiated mammary tumors. Thus, from the evidence currently available, blocking the PR would be predicted to result in anti-tumor effects in the breast.

Androgen receptor and cancer. The biological activities of endogenous androgens, the most important being 5 α-dihydrotestosterone, are mediated primarily by the androgen receptor.[7] The major source of endogenous androgens in the male is the testis and in the female is the adrenal cortex and the theca cells of the ovaries. Small, but perhaps significant, levels of androgens are also synthesized in the breast, bone, and brain.[8] In the male, the tissues expressing the highest levels of androgen receptor (AR) are those within the male reproductive tract, in particular the testis and within the luminal cells of the prostate.[9] Androgens and appropriate androgen receptor signaling are critical for the maintenance of homeostasis in the adult prostate epithelium. However, the activated androgen receptor pathway also plays a major role in promoting the development and growth of prostate cancers which arise primarily from the luminal epithelium of the prostate. Interestingly, mutations in the AR which frequently develop during prostate tumorigenesis render the receptor sensitive to the effects of 17β estradiol, thereby driving proliferation of the prostate tumor cells.[10] A role of the AR pathway in breast, ovarian, and endometrial cancers is also emerging as it is expressed in all stages of mammary development and its signaling impacts both uterine growth and ovarian function.[11] Like ERβ, the AR opposes ERα signaling and thus its activation is thought to protect against estrogen-mediated tumorigenesis. However, the anti-versus pro-tumorigenic actions of the AR may be tissue-dependent. For example, activation of the AR decreases proliferation of breast and endometrial tumor cells, but appears to enhance the growth of at least some ovarian tumor cells. In addition, AR influences other important tumorigenic pathways such as ERBB2/HER2, BRCA, WNT/β-catenin, PIK3CA/AKT, ERK, and PTEN.

Impact of flavonoids on steroid receptor signaling. Many studies have examined the ability of flavonoids to either activate or inhibit the estrogen receptors. For example, Collins-Burrow *et al.* identified biochanin A, chrysin, and genistein as the most potent with respect to their ability to activate ERα.[12] On the other hand, luteolin and to a lesser extent, kaempferide, flavone, and apigenin were found to harbor anti-estrogenic properties. The anti-estrogenic activities of some flavonoids (i.e., luteolin, kaempferol, quercidin, myricetin, and isorhamnetin) have been attributed to their ability to inhibit aromatase expression and thereby block estrogen biosynthesis.[13] The differing biological effects of flavonoids have also been attributed to their varying abilities to interact with ERα versus ERβ.[14] While 17β estradiol activates both receptors, some flavonoids, in particular genistein, S-equol, liquiritigenin, anddaidzein, exhibit ERβ selectively.

As compared to 17β estradiol, the activities of these so-called "botanical estrogens" are less stimulatory of genes that regulate proliferation and migration and more stimulatory of those involved in promoting apoptosis activities consistent with activation of ERβ.[15] It has been suggested that the ability of these botanical estrogens to induce cell proliferation is highly dependent on the relative expression levels of ERα versus ERβ.[16]

Flavonoids have also been found to interfere with the actions of the AR and its ability to facilitate testosterone-induced proliferation. For example, a screen of representative flavones, flavanones, and isoflavones identified 6-hydroxyflavanone as harboring one of the highest binding affinities for the AR, albeit at several orders of magnitude less than that exhibited by endogenous androgens.[17] Similarly, a structure activity screen that measured both binding to the AR and either enhancement or inhibition of AR-mediated gene transactivation identified 5-hydroxyflavone as an AR antagonist and 6-methoxyflavone, genistein, and daidzein as AR agonists.[18] However, genistein has also been shown to act as a partial agonist/antagonist of the AR when administered *in vivo*.[19] Whether genistein activates or inhibits the AR may depend on its tissue concentrations and the cellular milieu. For example, when administered in the presence of androgens, genistein's effects were found to be anti-androgenic in the testis, prostate, and brain. In the absence of androgens, however, genistein exerted agonistic effects in the prostate and brain. In addition to binding and blocking the AR, flavonoids may also inhibit upstream events that impinge on AR activity. For example, flavonoids such as isoliquiritigenin can block the activity of aromatase and in this manner, block androgen signaling.[20]

Impact of flavonoids on the efficacy of anti-steroid treatments. Studies questioning whether co-adminstration of flavonoids would impact the ability of anti-steroid therapies to inhibit tumorigenesis have generated mixed results. *In vivo* administration of genistein has been shown to abrogate the abilities of both the ERα antagonist, tamoxifen, and the aromatase inhibitor letrozole to inhibit the growth of explanted human breast cancer.[21,22] Tangeretin exerted similar effects.[23] A common problem observed in patients treated with anti-steroid therapies is that their tumors typically develop resistance. Using tamoxifen-resistant breast cancer cells, Tu *et al.* reported that luteolin could sensitize the cells to the growth inhibitory effects of tamoxifen.[24] Similarly, apigenin alone restored the ability of fulvestrant or tamoxifen to inhibit growth in cells that had become resistant to their respective treatments.[25] Co-treatment with apigenin and either fulvestrant or tamoxifen resulted in synergistic activities. A potential mechanism by which flavonoids may restore the cellular response to the anti-proliferative effects of anti-steroids is via epigenetic mechanisms.[26] For example, treatment of tamoxifen-resistant cells with apigenin reactivated the expression of ERα via epigenetic mechanisms and enhanced the ability of tamoxifen to block tumor growth. Thus, additional research is required to determine whether consumption of flavonoids is harmful or beneficial to patients undergoing anti-steroid cancer treatments. While flavonoids that exhibit selectivity toward activating ERβ or when administered in steroid-poor environments may be beneficial, the ultimate impact of flavonoid intake on therapeutic outcomes is yet to be determined.

Overview of breast cancer. Breast cancer is the most frequent type of cancer in women throughout the world.[27] The incidence rates reported in 2011 were found to vary from a low of 19.3 per 100,000 women in Eastern Africa to a high of 89.9 per 100,000 women in Western Europe. However, breast cancer incidences are dramatically increasing in developing countries and in particular, those with considerable economic development such as Japan, Singapore, and urban areas within China.[28] Age, as well as lifetime exposure to elevated levels of steroid hormones, are the major risk factors associated with breast cancer. In Asia, the peak years for developing breast cancer are between 40 and 50 years of age whereas in western countries, the peak occurs within 60 to 70 years of age.[27] Additional risk factors are delayed childbearing, lower parity, decreased breastfeeding, increased body mass index, and increased dietary fat consumption.[28] With respect to underlying genetic susceptibility, the best characterized "breast cancer genes" are BRCA1 and BRCA2 which are involved primarily in DNA repair. Loss of function mutations in these genes are highly penetrant, but relatively rare. In the United States, for example, the incidences of women who carry loss of function mutations in BRCA1 and BRCA2 are 1 in 300 women and 1 in 500 women, respectively.[29]

Breast cancer is a heterogenous disease that is typically classified into four intrinsic subtypes based on the expression levels of ERα and PR as well as overexpression of the growth hormone receptor ERBB2/HER2.[30] The subtypes described as "luminal A" and "luminal B" express both ERα and PR. Luminal A tumors are characterized as histologically low grade whereas luminal B tumors express lower levels of hormone receptor and are histologically high grade. The "ERBB2/HER2 enriched" subtype expresses very high levels of ERBB2/HER2 due to its over amplification and do not express either ERα or PR. Finally, the "triple negative" basal subtypes lack expression of ERα and PR and do not exhibit over-expression of ERBB2/HER2. Although epidemiological studies are as yet inconclusive, some data indicates that the majority of breast cancers express hormone receptors and that breast cancer in women with African ancestry may represent a higher proportion of the triple negative, basal subtype.[28]

In addition to ERs, other signaling pathways that play important roles in breast cancer development are TP53, PTEN, BRCA1/BRCA2 and ERBB2/HER2. As mentioned previously, BRCA1/BRCA2 are tumor suppressor genes involved in DNA repair.[31] The cytoplasmic form of BRCA1 regulates mitotic cell division, cytoskeletal rearrangements, apoptosis, and mitochondrial genome repair. During conditions of genotoxic stress and DNA damage, BRCA1 translocates to the nucleus where it regulates DNA damage repair processes, DNA replication, gene transactivation, and X chromosome inactivation. Mutational inactivation of BRCA1 leads to accelerated growth of breast and ovarian cells. Similarly, BRCA2 is important for maintaining genome stability and also plays an important role as a co-regulator of gene transcription. ERBB2/HER2 is a member of the tyrosine kinase receptors and is amplified in approximately 20% of breast cancers.[24,32] ERBB2/HER2 is activated not by ligand binding, but following heterodimerization with other activated family members or when overexpressed, by homodimerization. In addition, activated steroid receptors, such as the ER, can activate ERBB2/HER2 signaling via their non-genomic mechanisms. Activation of ERBB2/HER2 enhances the transcriptional activation of downstream transcription factors that are involved in cell proliferation, cell survival and differentiation, angiogenesis, and invasion and metastasis.

Chemoprevention of breast cancer. Evidence generated over the past several years including results obtained from 11 randomized clinical trials indicate that pharmacological inhibition of the estrogen receptor pathway (anti-estrogens) is effective in inhibiting breast cancer.[33] Participants in these studies were women who were identified to be at high risk for developing breast cancer due to their family histories. The anti-estrogens used included either selective estrogen receptor modulators (SERMS) such as tamoxifen and raloxifene or aromatase inhibitors (i.e., anastrozole and lestrozole). Tamoxifen is perhaps the best studied and its use over a five-year period has been shown to prevent approximately 50% of the ER-positive breast cancers. Raloxifene use is similarly effective with reductions of 44% to 65% observed. However, use of both drugs is accompanied by increased susceptibility of developing venous thrombosis. Further, use of tamoxifen, but not raloxifene, is associated with increased incidences of endometrial cancers. This is due to the fact that the actions of tamoxifen are tissue-dependent. In the breast, it acts as an ER antagonist, but in the endometrium, it exerts ER agonist activities. The use of aromatase inhibitors to prevent breast cancer is less well studied. From the data generated thus far, it appears that they are more effective than tamoxifen in preventing breast cancer in post-menopausal women. Tamoxifen is still recommended for pre-menopausal women. Other proposed pharmacological agents to be used for chemopreventive approaches, in particular to prevent estrogen receptor negative breast cancers, include statins and NSAIDS.[34,35] In addition, changes in lifestyle choices are often recommended as preventative approaches and include maintaining proper weight and/or body mass index, increasing the consumption of plant-based foods, increasing physical activity, and decreasing alcohol consumption.[33] The potential of flavonoids as indicated by pre-clinical, clinical intervention, and epidemiological studies will be discussed in the following sections.

Flavonoids and breast cancer. Flavonoids that have been shown to inhibit the proliferation and/or induce apoptosis of cultured mammary tumor cells include apigenin,[25] fisetin,[36] isoliquiritigenin,[37] luteolin,[38] morin,[39] naringin,[40] quercetin,[41] silibinin,[42] tricetin,[43] triticuside A,[44] and wogonin.[45] In addition, metabolites of both quercetin and catechin harbor anti-proliferative activities.[46] Further, protoapigenone, a natural derivative of apigenin, has been found to be more potent than apigenin with respect to their ability to inhibit *in vitro* growth.[47] Mechanisms proposed to underlie these events include activation of the caspase cascade (fisetin,[36] luteolin,[38] silibinin,[42] tricetin[43]), inhibition of pro-survival pathways such as PIK3CA and AKT (isoliquiritigenin,[37] morin,[39] and wogonin[45]), inhibition of β-catenin (naringin[40]), upregulation of ERK (wogonin[45]) and inhibition of proteosomal activity (apigenin[48]). Interestingly, a panel of flavonoids was found to exhibit varying abilities in inhibiting FAS activity, lipolysis, and subsequently, apoptosis.[49] Some studies have shown that flavonoids act in a biphasic manner. For example, at lower concentrations (~1 µM) apigenin, genistein, and quercetin increased proliferation presumably via their activation ERα.[25,50] However, at higher concentrations (greater than 10 µM), they inhibited the actions of pro-survival proteins, such as MAPK and AKT, and inhibited growth.

Several studies have shown that *in vivo* administration of flavonoids such as apigenin,[48] biochanin A, quercetin, and EGCG[51] can inhibit the growth of explanted mammary tumor cells. Chemopreventive effects have also been shown using a two-stage carcinogen-induced model of mammary tumor formation. Here, the mice were first administered the carcinogen (DMBA) and were then administered medroxy progesterone acetate to initiate and promote mammary carcinogenesis, respectively.[52,53] Apigenin was administered throughout the course of the experiment. When administered by intra-peritoneal injections, apigenin inhibited the growth and development of the tumors.[52] When administered via the diet, however, apigenin decreased tumor incidence.[53] Within the mice that did develop tumors in this latter study, more tumors were formed indicating that apigenin had increased tumor multiplicity. Studies using silibinin have shown that the ability of silibinin to inhibit the growth of explanted human tumors requires an intact host immune system.[54]

In addition to inhibiting growth and inducing apoptosis, flavonoids such as glabridin,[55] isoliquiritigenin,[56] morin,[39] and baicalein,[57] have been shown to decrease migration and invasion. Further, *in vivo* administration of either chrysin[58] or isoliquiritigenin[56] inhibited metastasis. Finally, isoliquiritigenin[59] andglabridin[55] have been shown to inhibit expression of VEGF and angiogenesis. Interestingly the concentration of glabridin used to inhibit markers of angiogenesis had no impact on cell viability.

Given that a number of flavonoids can interact with ERα and ERβ with varying abilities and the key roles of these receptors in breast cancer development, the extent to which human consumption of flavonoids impacts the risk of developing breast cancer has been of high interest, but as yet, is not particularly definitive. The inconclusive nature of these studies is thought to arise from the complexity of breast cancer as a disease state; in particular, whether the participants are pre-menopausal or post-menopausal, and the molecular events involved in their tumor formation and progression. The most recent epidemiological studies that have examined the relationship between flavonoid intake and breast cancer incidence include a meta analysis that assessed 12 case control or cohort studies involving 9,513 breast cancer cases and 181,906 controls and encompassing a 13-year period.[60] The involved populations resided within the United States, Finland, China, the Netherlands, Mexico, Italy, and Greece. A high overall intake of flavonols and flavones was found to correlate with significant decreases in breast cancer incidences in post-but not pre-menopausal women. Similarly, a case control study of Korean women with 358 confirmed breast cancer cases and 360 controls found a protective effect of isoflavone intake (isoflavone intake ranged from 15 to 76.5 mg/day) only in post-menopausal women with ER positive breast cancer.[61] Finally, a population-based cohort study found that soy as well as isoflavone intake was associated with reduced risk of breast cancer.[62] The population examined in this group resided in Takayama City, Japan, and consisted of 17,125 of which 187 were diagnosed with breast cancer after 16 years. Isoflavone intake varied from a mean of 19.9 to 67.4 mg/day.

Other recent studies have failed to identify a protective link between flavonoid intake and breast cancer risk. Zamora-Ros *et al.* reported on results from the EPIC (European Prospective Investigation into Cancer and Nutrition) study.[63] This large prospective study involved 334, 850 women residing in Denmark, France, Germany, Greece, Italy, the Netherlands, Norway, Spain, Sweden, and the United Kingdom. The consumption of the six classes of flavonoids was monitored and total flavonoid intake ranged from less than 176 to more than 654 mg/day. After approximately 11.5 years, 11,576 breast cancers were diagnosed. No association between intake of total flavonoids or any flavonoid subclass and breast cancer risk was found, regardless of whether the participants were pre-or post-menopausal or whether their breast cancers were ER positive or negative. Similarly, a prospective cohort of 56, 630 women who resided in 21 states in the United States was performed.[64] The consumption of seven classes of flavonoids, including proanthocyanidins, was monitored. Total flavonoid intake ranged from less than 199 mg/day to more than 2,063 mg/day. There was no association between total flavonoid intake and overall breast cancer risk regardless of whether the breast tumors were ER positive or ER negative. However, women with higher consumption of flavan-3-ols were at lower risk of developing ER negative breast cancers. A nested case control study (Shanghai Women's Health Study) that compared urine levels of polyphenols and breast cancer risk has also been performed.[65] In this study, the plasma levels of quercetin, kaempferol, epicatechin, EGCG, and their respective major metabolites was measured in 353 patients diagnosed with breast cancer to that of 701 matched controls. The average urinary excretion of total polyphenols ranged from 21 to 33 ng/mg creatinine. No association between breast cancer incidence and the urine levels of any polyphenol was found. Finally, a population-based observational study of women residing in California and Hawaii was reported.[66] This multiethnic study examined populations of Japanese American, white, Latina, African American, and Native Hawaiian women. During the 12.5-year study, 4,769 of the 84,550 women developed breast cancer. Dietary isoflavone intake estimated from the questionnaires ranged from 0 to 178.7 mg/day. No statistical association between isoflavone intake and breast cancer risk was found.

Flavonoids and breast cancer chemoprevention. In addition to the epidemiological studies examining the impact of soy flavonoids on the risk of developing breast cancer, a few clinical intervention studies have been performed. A recently reported study used a randomized placebo controlled design to address the question of whether high levels of soy supplements were beneficial or harmful to women with respect to breast cancer progression.[67] Here, 140 women who were diagnosed with invasive breast adenocarcinomas were given 25.8 grams of either soy or milk protein for 14–15 days. Analyses of their plasma levels revealed that those receiving soy protein had genistein levels that ranged from a median of 1.6 to 11.6 ng/ml and daidzein levels of 1.5 to 6.7 ng/ml. This represented an increase of seven-fold and four-fold of their initial plasma levels of genistein and daidzein, respectively. The plasma levels of genistein and daidzein were not significantly altered in the placebo controls. In the patients with high genistein plasma levels, increases in the expression levels of genes associated with tumor growth were observed.

The potential impact of isoflavone consumption on the effectiveness of anti-estrogen therapy has also been examined.[68] Here, 524 women who were receiving hormone therapy consisting of either the ER antagonist tamoxifen or the aromatase inhibitor anastrozole were studied. The patients were 29 to 72 years of age and were diagnosed with ER positive breast cancer. Of those receiving tamoxifen, 248 were pre-menopausal and 190 were post-menopausal. Only post-menopausal women (86 patients) were treated with anastrozole. During this five-year study, the mean daily intake of isoflavones was 25.6 mg/day and was primarily from soy milk, tofu, and soy flour. Amongst the pre-menopausal women, soy isoflavones did not appear to impact either breast cancer recurrence or death. However, amongst the post-menopausal women, high consumption levels of soy isoflavones corresponded to a decreased risk of breast cancer recurrence.

Taken together, the studies described above indicate that isoflavones may have an impact on breast cancer risk and tumor development. It has been recently suggested that the issue of whether consumption of high levels of flavonoids that harbor estrogenic activities, such as genistein and diadzein, increases or decreases a woman's risk of developing breast cancer may be dependent on whether or not the breast cancer cells exist in an estrogen-rich versus an estrogen-poor environment.[69] Here, an estrogen rich environment exists within the breast tissue of a pre-menopausal woman whereas in a post-menopausal woman, the environment of the tissue is estrogen-poor. During the onset of menopause, the breast cancer cells are thought to adapt to their increasingly estrogen-poor environment by acquiring a hypersensitivity to estrogen. This makes them vulnerable to estrogen-induced apoptosis. An estrogen-poor environment also exists in breast cancer patients who have undergone extensive anti-hormone therapy. Here, administration of estrogen at high doses can induce apoptosis and hence, remission of breast cancer. Given this scenario, consumption of high levels of soy prior to menopause may enhance the pro-proliferative effects of estrogen in the estrogen-rich environment thereby enhancing breast cancer incidence. After menopause and in the context of an estrogen-poor environment, consumption of high levels of phytoestrogens that are sufficient for driving estrogen receptor-mediated apoptosis may prove to be beneficial. The potential impact of flavonoid consumption on the growth and progression of breast tumors in the pre- versus post-menopausal woman is shown in Figure 5.1.

In addition to soy flavonoids, other flavonoids that have been investigated with respect to their impact on breast cancer and biomarkers

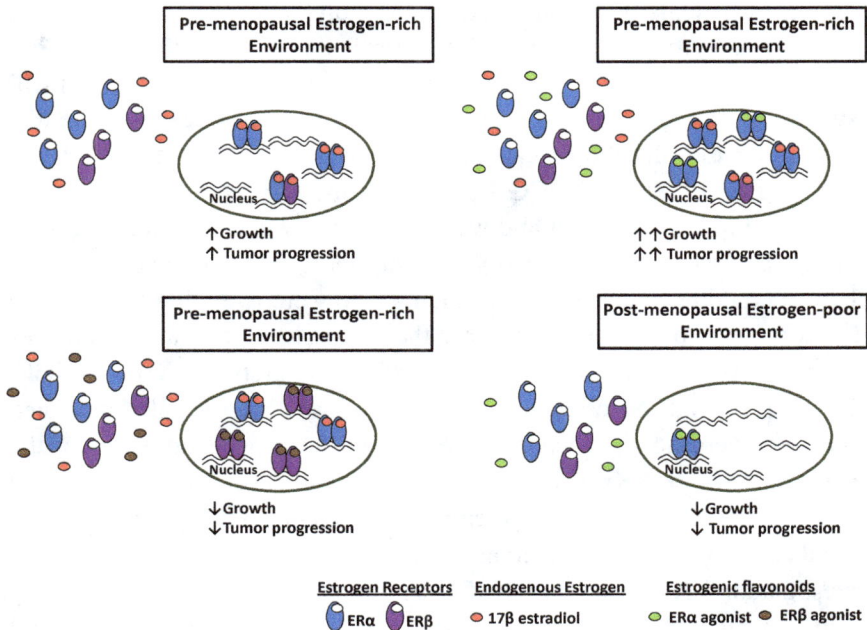

Figure 5.1. Putative impact of flavonoids on breast tumor cells. The events that maybe initiated by flavonoids with estrogenic activity ("Estrogenic Flavonoids") in the breast tissue of a pre-versus post-menopausal woman are depicted. In the estrogen-rich environment of the pre-menopausal woman, ERα is activated by both endogenous estrogen, 17β estradiol, and estrogenic flavonoids, such as genistein. Here, their activation of ERα enhances the growth and progression of the breast tumor cells. However, flavonoids that selectively activate ERβ may exert growth inhibitory effects and slow the progression of the breast tumor cell. In the estrogenic poor environment of the post-menopausal woman, however, activation of ERα is thought to induce apoptosis and induce breast cancer remission. Thus, exposure to estrogenic flavonoids that act as ERα in this scenario may prove to be beneficial.

associated with breast cancer include the tea catechins. Wu *et al.* used Polyphenon E® capsules to administer either 400 or 800 mg of EGCG to healthy post-menopausal women for two months.[70] A total of 103 individuals participated in this study. As compared to the placebo controls, the urinary levels of tea catechins were higher in the Polyphenon E®-treated patients, but there were no significant differences in the serum levels of estrogen or of measures of liver function. In addition, body weight and body mass index was unchanged. However, serum levels of low-density lipid cholesterol were significantly decreased in the Polyphenon E®-treated patients. A second study using Polyphenon E® and a phase I, randomized double-blinded, placebo-controlled dose escalation design has also been reported.[71] Forty patients who were diagnosed with stage I–III, estrogen and progesterone receptor negative breast cancer were administered either the placebo or capsules to deliver a daily dose of 800 mg, 1,200 mg or 1,600 mg EGCG for six months. Adverse effects were not considered to be significant and were primarily associated with GI disturbances (nausea, diarrhea, constipation, and indigestion).Urinary tea polyphenols could be detected in the patients who were receiving Polyphenon E®. Compared to the placebo control, there were no significant differences in serum levels of estradiol, sex hormone binding globulin, IGF-1 or IGF-binding protein 3. Assessment of the breast tissue biopsies also indicated that there were no significant differences in the expression levels of the proliferation marker Ki-67 between the placebo and Polyphenon E®-treated groups. Finally, the impact of green tea capsules in women diagnosed with ductal carcinoma in situ or primary invasive stage I or II breast cancer was investigated.[72] The amount of EGCG administered per day was estimated to be 940 mg. After 35 days, the urinary levels of tea catechins were found to be increased by 2–10-fold (*n*=13) and were essentially unchanged in the placebo control (*n*=18). The expression levels of biomarkers representative of proliferation (Ki-67), angiogenesis (CD34) and apoptosis (caspase-3) were evaluated in breast biopsy samples obtained prior to and following tea administration. Of these, expression levels Ki-67 were significantly decreased in the benign tumor cells in the tea versus placebo-treated patients. The expression levels of CD34 and caspase-3 were unchanged. Overall, the results obtained from these EGCG clinical intervention studies are promising, but inconclusive, in part due to the small number of subjects, the uncertainties regarding the appropriate dose of EGCG required to elicit a favorable response, and the identification of the patients that would be most responsive to the effects of EGCG.

Overview of ovarian cancer. Worldwide, ovarian cancer is the sixth most commonly diagnosed cancer representing approximately 6.6 new cases/100,000 women each year.[73] Because of its asymptomatic nature, it is frequently diagnosed at relatively advanced stages, stages III or IV. If diagnosis occurs when the disease is localized, the five-year survival rates are 100% but drop to 28% when diagnosis occurs at the metastatic stage of the disease. The majority (90%) originate from epithelial cells with the remaining originating from germ cells, sex cords, and ovarian stromal cells. The source of epithelial-derived ovarian carcinomas is yet to be established, but is postulated to arise from the Fallopian tubes. Ovarian carcinomas are heterogeneous and composed of five classes: low grade serous carcinoma, high grade serous carcinoma, mucinous carcinoma, and endometriod carcinoma. At the molecular level, low grade serous carcinomas often harbor KRAS and BRAF mutations whereas high grade serous carcinomas often express mutated TP53. Mutations frequently observed in endometriod carcinomas include ARID1A (a component of the SWI-SNF chromatin remodeling complex), β-catenin and PTEN (a protein tyrosine phosphate family member). The most frequently identified mutations observed in clear cell carcinomas are those within the ARID1A, PTEN, KRAS, and PIK3CA genes. In addition, mutated forms of BRCA1 and BRCA2 play important roles in hereditary ovarian cancer and increase the risk of developing ovarian cancer from 1% to 27–44%.

While genetics play an important role in determining a woman's risk of developing ovarian cancer, other involved risk factors include early menarche, late menopause, delayed childbearing (i.e., continuous ovulation), smoking, obesity, and high-fat diets. The incidence of ovarian cancer is highest in white women of mid to high socio-economic status. It also occurs more frequently in patients diagnosed with breast cancer-ovary syndrome, Li-Fraumeni syndrome, Lynch syndrome, and polycystic ovaries. Patients with Lynch syndrome are prone to developing hereditary nonpolyposis colorectal cancer and harbor mutations in DNA mismatch repair genes. Polycystic ovary syndrome arises from an increase in systemic androgen levels that interferes with the ovarian-steroid feedback loop. As a result, follicle-stimulating hormone levels increase and induce hyperactivity within the stromal and thecal cells of the ovaries. With respect to steroid hormone receptors, studies indicate that expression of PR is associated with a longer disease-free survival in patients with endo-metriod carcinomas and high grade serous carcinomas. Expression of ER correlates with longer disease-free survival in patients with endometeroid carcinomas.

Chemoprevention of ovarian cancer. The results of several large population-based studies indicate that the use of oral contraceptives decreases the risk of developing ovarian cancer.[74] Unlike endometrial cancer, which is protected primarily by progestin-containing formulations, the protection of ovarian cancer does not appear to require the presence of progestin and is independent of formulation. The reduction of cancer risk persists for at least 30 years after the discontinuation of oral contraception and is effective in women with BRCA mutations. Reduced risk of ovarian, particularly the mucinous and endometriod types, has also been observed in women who take low-dose aspirin.[75,76] Because of our lack of understanding of the etiology and progression of ovarian cancer, animal models of ovarian disease and pre-clinical preventative studies are lacking. Thus, our understanding of how flavonoid intake may impact ovarian carcinogenesis is limited to studies performed using cultured ovarian tumor cells, tumor explants and epidemiology studies.

Impact of flavonoids on ovarian carcinogenesis. Flavonoids that have been shown to inhibit proliferation and/or induce apoptosis of cultured ovarian carcinoma cells include apigenin,[77] kaempferol,[78] protoapigenone,[79] silibinin,[80] silymarin,[81] and trifolirhizin.[82] It is important to note that in some cases (i.e., kaempferol[78] and trifolirhizin[82]), very high levels, 40–50 uM, were required to initiate significant inhibition of growth and apoptosis. These doses are physiologically unrealistic. Proposed underlying mechanisms include activation of TP53, BAX, and caspases,[81] enhanced formation of reactive oxygen species, and downregulation of the ERK/AKT survival pathway.[80] In addition, quercetin was found to enhance apoptosis initiated by TRIAL/TNFSF10 via mechanisms that involve the JNK pathway. Apigenin has been shown to reduce the ability of ovarian tumor cells to adhere, migrate, and invade the synthetic matrix *in vitro*.[83] In addition, apigenin can inhibit induction of VEGF expression by hypoxia-inducing factor alpha and inhibit *ex vivo* angiogenesis by modulating the PIK3CA/AKT pathway.[84] *In vivo* explant studies have shown that silibinin and protoapigenone can inhibit ovarian tumor growth.[79,80] Further, apigenin inhibits metastasis of explanted ovarian tumors via mechanisms that may involve the PIK3CA/AKT/mTOR and FAK pathways.[85,83] However, while *in vivo* treatment with quercetin slightly inhibited growth of explanted ovarian tumors, its co-administration with the chemotherapeutic drug cisplatin resulted in a tumor-promoting effect.[86]

Human population studies that have examined the impact of dietary intake of flavonoids on the incidence of ovarian cancer include a study of 38,408 female health professionals who resided in the United States and were 45 years of age and older.[87] Total dietary flavonoid intake of these women ranged from 8.88 to 47.44 mg/day. The greatest contributor of total flavonoid intake was quercetin, followed by kaempferol, myricetin, apigenin, and luteolin. No association between flavonoid intake and incidence of ovarian intake was observed in this 11.5-year follow-up study. A case control study of patients who resided in either Massachusetts or New Hampshire was also reported.[88] Here, 1,141 cases of ovarian cancer and 1,183 controls were studied. Total dietary flavonoid intake ranged from 0.9 to 95 mg/day in these subjects. The authors reported no evidence of association between total flavonoid intake and risk of developing ovarian cancer. However, intake of apigenin was reported as borderline significant with respect to reduced ovarian cancer risk. A second case control study examined an Australian population.[89] This study consisted of 1,366 cases of ovarian cancer and 1,414 control subjects. The median intake of total phytoestrogens was 1.2–1.3 mg/day with isoflavone intake representing 0.5–0.6 mg/day. No association between flavonoid intake and ovarian cancer risk was observed. Finally, a clinical study was performed in patients with advanced ovarian cancer.[90] Here, five of the 16 patients who were administered enriched green tea remained free of ovarian cancer occurrence.

Overview of endometrial cancer. Endometrial cancers are the sixth most common cancers in women worldwide with higher rates in developed countries (approximately 12.9/100,000 women) as compared to those in developing countries.[91–93] These cancers typically occur in women over the age of 40 and are usually detected at stage I due to its presentation of vaginal bleeding. When diagnosed at stage I, five-year survival rates are approximately 90%, but are significantly decreased when the disease has progressed to more advanced stages with metastasis to the ovaries, abdomen, and lymph nodes. The majority of endometrial cancers are adenocarcinomas which are subdivided into type I and type II. Risk factors for developing endometrial cancer are similar to those of the breast and ovary. Because unopposed estrogens cause uterine hyperplasia, high estrogen levels are thought to be involved in the development and progression of uterine cancer. Progestins, however, induce differentiation of the endometrial epithelial cells and are the most common treatments used for uterine cancer. It is also thought that ERβ plays an important role in maintaining uterine homeostasis and that its aberrant signaling is involved in the carcinogenic process. Despite the apparent role of estrogens in endometrial carcinogenesis, use of anti-estrogens as a therapy has met with limited success. Genetic predispositions include hereditary nonpolyposis colon cancer which involves defects in the DNA mismatch repair pathway. In addition, type I endometrial tumors often harbor mutations in PTEN, KRAS, and β-catenin as well as microsatellite instability. Type II tumors are estrogen-independent and often contain mutations in TP53, inactivation of the CDKN2A/p16 tumor suppressor, and amplification of ERBB2/HER2. Laboratory animal models used to study endometrial cancer include mice that lack uterine expression of PTEN or both PTEN and STK11/LKB1, a negative regulator of the mTOR pathway.[93,94]

Chemoprevention of endometrial cancer. The observation that progestin-containing oral contraceptives was associated with a decrease in the risk of endometrial cancers in the general population led to a phase II randomized clinical trial involving 51 women who had Lynch syndrome.[95] These patients were either confirmed as harboring mutations of DNA repair genes (MLH1, MSH2, or MSH6) and/or had a family history of endometrial cancer. They received either 0.3 mg ethinyl estradiol

+ 0.3 mg norgestrel daily or 150 mg depomedroxy progesterone acetate for 21 days. Endometrial biopsies obtained prior to and following treatment revealed that in both treatment groups, a significant decrease in proliferation (as indicated by the number of cells expressing the marker Ki-67) was observed. In addition, decreases in the expression levels of IGF-1 and the apoptosis inhibitor survivin were detected. Additional chemopreventive interventions under consideration for inhibiting endometrial cancer include NSAIDs and metformin. Use of aspirin, but not NSAIDs (ibuprofen and naproxen), has been associated with a decreased risk of developing endometrial cancer.[96] With respect to metformin, two clinical trials examining the impact of metformin on endometrial cancer were recently reported.[97,98,107] In the first trial, 20 obese women who were diagnosed with endometrioid cancer were administered metformin (8,000 mg per day) for up to four weeks.[97] At the end of the trial, tumor biopsies indicated that metformin treatment reduced tissue levels of the proliferation marker Ki-67 as well as members of the mTOR signaling pathway. Further, the expression levels of ER, but not PR, were reduced. In the second trial, 43 women with abnormal uterine bleeding and either disordered proliferative endometrium or simple hyperplasia were involved and were randomly assigned to receive either metformin (500–1000 mg) or medroxyprogesterone acetate-Megestrole (40 mg daily) for three months.[98] Comparison of uterine biopsies obtained prior to and following treatment indicated that 21 of the 22 patients receiving metformin and 13 of the 21 patients receiving medroxyprogesterone acetate-Megestrole indicated a positive response which was histologically defined as atrophic endometrium.

Impact of flavonoids on endometrial carcinogenesis. Flavonoids that have been shown to inhibit proliferation and/or induce apoptosis in cultured endometrial cancer cells include EGCG,[99] eupatilin,[100] and icaritin.[101] The underlying mechanisms of EGCG's actions are thought to involve upregulation of p38 kinase, downregulation of ERK kinase signaling and enhanced formation of reactive oxygen species and oxidative stress.[99] Eupatilin was found to induce G2/M arrest, exhibited relatively low cytotoxicity, and inhibited the activity of the mutated p53 suppressor.[100] Icaritin is thought to induce apoptosis via sustained activation of

ERK1/2 signaling.[101] Other models used to probe the impact of flavonoids using *in vitro* approaches include co-culturing adenocarcinoma cells with stromal cells.[102] Here, genistein was found to exert a biphasic effect on estrogen-induced growth of the adenocarcinoma cells. At low concentrations, genistein blocked estrogen-induced proliferation, but at higher concentrations, estrogen-induced proliferation was enhanced. The mechanisms proposed to underlie genistein's effects are activation of ERβ at low concentrations (anti-proliferation) and activation of ERα at higher concentrations (pro-proliferation).Using *in vivo* approaches, daidzein was not found to enhance cell hypertrophy in the rat uterus, but at the concentration used, was capable of enhancing ERα activity.[103] Genistein, however, inhibited estrogen-induced cell proliferation in the rat uterus.[104] In addition, the administration of extracts containing isoflavone aglycones enhanced uterine proliferation and when co-administered with the carcinogen DMBA increased the formation of atypical, hyperplastic tissue and adenomatous polyps in the endometrium.[105]

Epidemiological studies examining the relationship between flavonoid intake and endometrial cancer include the aforementioned case control Australian study which consisted of 1,288 endometrial cancer cases and 1,435 control subjects.[89] The median intake of total phytoestrogens was 1.3 mg/day of which 0.6 mg/day was isoflavones. The authors found no evidence of association between intake of phytoestrogens or of isoflavones and risk of endometrial cancer. An Italian case control study that consisted of 454 women with endometrial cancer and 908 controls was also performed.[106] The calculated median intake of flavonoids (mg/day) was as follows: flavanols, 36.6; flavanones, 32.3; flavonols, 18.7; anthocyanidins, 0.5. The only significant correlation observed was an inverse association between high dietary intake of proanthocyanidins and the risk of endometrial cancers. Quaas *et al.* performed a double-blind placebo controlled trial to determine the impact of isoflavone soy supplements on the risk of developing endometrial cancer.[107] The trial consisted of 350 post-menopausal women who were administered a supplement consisting of either 25 g soy protein (containing 154 mg of total isoflavone conjugates and aglycones) or 25 g milk protein for three years. The average plasma levels of daidzein, genistein, and glycitein were 290.5, 354, and 10.4 nmol/l, respectively, in the soy-administered group as compared to

41.5, 27.1, and 3.5 nmol/l in the placebo group. While the rate of uterine hyperplasia/malignancy appeared to be higher in the placebo group as compared to that of the soy supplemented group, this difference was not statistically significant.

A prospective analysis of 46,027 post-menopausal women residing in either Hawaii or California was also performed to determine the impact of isoflavone intake on the risk of endometrial cancer.[108] The energy-adjusted intake of total isoflavone ranged from less than 1.59 mg/1,000 kcal per day to more than 7.82 mg/1,000 kcal per day. A significant decrease in the risk of endometrial cancer was associated with the highest intakes of total isoflavones, daidzen, and genistein.

Overview of prostate cancer. Worldwide, prostate cancer is the sixth leading cause of deaths and the second most frequently diagnosed cancer in males.[9,109] Age-related declines in plasma androgen levels correspond with the initiation of human prostate cancers. Thus, there is a paradoxyl relationship between serum androgen levels and prostate cancer risk. The underlying mechanisms are unclear, but may involve adaptation of benign tumor tissue within the prostate to the low serum testosterone levels that facilitates a permissive growth environment. The most effective therapy used to treat prostate cancer is androgen deprivation therapy, which includes castration or the administration of anti-androgens. Castration-resistant prostate cancers develop in some cases and are thought to harbor intratumor levels of androgens that continue to drive its growth and progression. It is not yet clear whether the AR is involved in the initiation of prostate carcinogenesis or whether it acts primarily as a tumor promoter via its inappropriate, sustained signaling that occurs due to mutational events. A number of polymorphisms of genes within the androgen pathway are associated with determining prostate cancer development including the AR itself as well as those involved in androgen biosynthesis.

Additional players in prostate carcinogenesis include members of the PIK3CA/AKT/mTOR pathway which are thought to participate in a reciprocal feedback mechanism with the AR which allows for continued AR signaling. Estrogen is also thought to contribute to prostate carcinogenesis. Loss of ERβ is associated with the progression of the normal tissue to prostate tumor tissue and activation of ERβ can inhibit the proliferative activity within the tumor tissue. Conversely, ERα is thought to be involved in the pro-inflammatory conditions that are associated with prostate tumor progression. Additional pathways thought to play key roles in the progression of prostate cancer are WNT/β-catenin, MET (a receptor tyrosine kinase, also called hepatocyte growth factor receptor), VEGF, and hepin. The WNT/β-catenin pathway regulates genes involved in proliferation (i.e., MYC), angiogenesis (i.e., VEGF), and migration/invasion (i.e., MMPs). MET is typically downregulated by the AR and is upregulated by RANKL (receptor-activator of NFκB). Hepin is a protease that cleaves and thereby activates peptides involved in growth factor signaling, such as pro-hepatocyte growth factor.

Animal models typically used to study prostate carcinogenesis include the TRAMP, PTEN, and MYC mouse models.[110] The TRAMP mouse was developed by overexpressing the viral oncogenic proteins (the large T and small T antigens of the SV 40 virus) in the prostate tissue. While it recapitulates many events associated with androgen-induced tumor progression, the tumors are classified as neuroendocrine, rather than the adenocarcinomas typically observed in human patients. In the PTEN mouse model, the tumor suppressor protein PTEN which is often deleted in human prostate cancers is lacking within the murine prostate tissue. These mice develop both intraepithelial neoplasia (PIN) and adenocarcinomas. An additional animal model that is used to study the progression of prostate cancer is mice that harbor overexpression of the MYC oncogene within the prostate tissue. These mice develop PIN, adenocarcinomas, and invasive prostate cancers.

Chemoprevention of prostate cancer. The understanding that testosterone played a key role in the development and progression of prostate cancers led to the development of chemopreventive approaches that have largely focused on inhibiting testosterone signaling.[111,112] Those proven to be most effective inhibit 5α-reductase which blocks the conversion of testosterone to its physiologically active form. Two forms of 5α-reductase exist. Type I is expressed in the liver and skin whereas type II is expressed in the prostate. Two pharmacological agents that inhibit 5α-reductase, finasteride and dutasteride, have been developed and tested in randomized clinical trials. Finasteride inhibits the type II form of 5α-reductase whereas dutasteride inhibits both forms. The results of two clinical trials, Prostate Cancer Prevention Trial and Reduction by Dutasteride of Prostate Cancer Events, indicated that both drugs were effective at reducing the risk of developing prostate cancer. However, apparent increases in the incidences of high-grade prostate cancer raised significant concerns. To address this concern, an 18-year follow-up of the finasteride trial was performed using mortality as an estimate of the presence of high-grade, fatal prostate cancers in all participants.[113] This study revealed that the overall rates of survival or survival after prostate cancer diagnosis were not significantly different when comparing the finasteride- versus placebo-treated groups. Despite these reassuring results, the safety associated with the use of 5α-reductase inhibitors remains under considerable discussion. It has been suggested that further clinical trials should proceed but with additional screening to identify individuals who are most likely to benefit from these therapies.[112] Additional chemopreventive agents that have been tested in randomized, placebo-controlled clinical trials include antioxidants.[114] The cumulative results of these trials indicate that there is no substantial evidence of a beneficial effect with respect to prostate cancer, arising from supplementation with folate, β-carotene, vitamin C, or vitamin D. In some cases, and as discussed in Chapter 1, supplementation with vitamin E and selenium has yielded conflicting results that have raised the possibility of dose-dependent relationships. Other agents that have shown promise in pre-clinical studies and may impact prostate-specific antigen (PSA) levels in patients include interventions with pomegranate, broccoli, turmeric, lycopene, and green tea.

Impact of flavonoids on prostate carcinogenesis. Flavonoids that have been shown to decrease the growth and/or increase apoptosis of cultured human prostate cancer cells include acacetin,[115] ampelopsin,[116] apigenin,[117,118] baicalein,[119] EGCG,[120] fisetin,[121,122] hesperidin,[123] isoliquiritigenin,[124,125] luteolin,[126] maysin,[127] nobiletin,[128] protoapigenone,[79] quercetin,[129] and vicenin-2.[130] The mechanisms by which these flavonoids have been proposed to exert their growth inhibitory actions include inhibition of pro-survival pathways such as AKT (apigenin,[117] fistin,[122] luteolin,[126] maysin,[127] nobiletin,[128] and vicenin-2[130]), mTOR (fisetin),[121] and p38/MAPK (acacetin[115] and apigenin[131]). Many of these studies have been performed in different prostate cancer cells that vary with respect to androgen receptor signaling to determine the extent to which flavonoids may inhibit the pro-proliferative effects of androgens. The LNCaP cell line expresses a mutated form of the AR and exhibit androgen-dependent growth. The growth of the PC-3 and DU145 prostate cancer cells, however, are androgen-independent. The growth-inhibitory effects exerted by flavonoids have in some cases been shown to be cell-line dependent. For example, treatment of LNCaP with either hesperidin or luteolin inhibited testosterone-induced growth of LNCaP cells, but did not impact the growth of PC-3 or DU145 cells, indicating that their effects on growth are likely mediated by androgen signaling.[123,132] Interestingly, genistein has been found to inhibit proliferation of prostate cancer cells that express the wild-type AR.[133] However, in cells expressing a mutant AR, genistein exhibited a biphasic effect and enhanced proliferation at low concentrations, but at higher concentrations inhibited proliferation. Fisetin also displayed cell-dependent activities where its ability to inhibit growth was more pronounced in the DU145 cell line as compared to that in the PC-3 cells.[121] Further, the ability of fisetin to inhibit AKT phosphorylation was observed only in the DU145 cells. However, vicenin-2 inhibited proliferation and induced apoptosis in both androgen-dependent and androgen-independent cell lines and was more potent than luteolin and orientin.[130] Given that stem cells within a tumor are thought to play important roles in the survival and progression of tumor cells, these cells may be critical targets of chemopreventive agents. With this in mind, Tang *et al.* isolated stem cells from the PC-3 and LNCaP cell lines and examined the impact of EGCG on their ability to self-renew.[134] EGCG was shown to not only inhibit self-renewal, but also to induce apoptosis and appeared to inhibit their ability to undergo epithelial mesenchymal transition. Co-treatment with quercetin enhanced these effects of EGCG.

A number of studies have attempted to clarify the mechanisms by which flavonoids may inhibit the growth and induce apoptosis of prostate cancer cells by analyzing the relative potencies of structurally similar compounds. Using the PC-3 cell line, the ability of four flavonoids (myricetin, myrcitrin, quercetin and quercitrin) to induce apoptosis was compared.[135] Of these, myricetin was found to be the most potent whereas quercitrin did not impact viability/apoptosis even at the highest concentration tested (600 μM). Similarly, the ability of linarin, linarin acetate, and acacetin to inhibit growth and induce apoptosis of LnCaP and DU145 cells was assessed.[136] Of these, the most potent was found to be acacetin. Finally, in the androgen-independent PC-3 cell line, the most potent with respect to growth inhibitor properties were quercetin, kaempferol, and luteolin and the least potent were naringenin, rutin, and genistein. Apigenin and myricetin were moderately potent. Other approaches used include a proteomics approach which identified vinculin, EF-1α, and hnRNPA1 as direct targets of quercetin in cultured prostate tumor cells.[137] The extent to which these proteins mediate the known biological properties of quercetin is yet to be determined.

In addition to modifying growth and apoptosis, flavonoids have been shown to decrease *in vitro* angiogenesis (luteolin[126] and vicenin-2[130]) and inhibit *in vitro* events associated with metastasis such as adhesion, migration and invasion (vicenin-2,[130] acacetin,[115] luteolin,[138] and EGCG[120]). The anti-angiogenic activities of flavonoids are thought to be mediated via their inhibition of VEGF[130] whereas their inhibition of migration and invasion is thought to involve inhibition of the EGFR pathway (i.e., quercetin[139]) as well as the activities of MMP2 and MMP9.[120,129] Other anti-cancer effects of flavonoids examined in cultured prostate cancer cells include induction of base-excision DNA repair.[140]

In vivo approaches that have been used to assess the impact of flavonoids on prostate tumorigenesis include the use of tumor explants, carcinogen-induced tumor formation, and the TRAMP mouse model. The growth of explanted prostate tumors has been inhibited by apigenin,[117] baicalein,[119] fisetin,[141] genistein,[142] luteolin,[126,132] protoapigenione,[143] silibinin,[144] and vicenin-2.[130] Further, treatment with luteolin inhibited the ability of the explanted prostate cells to metastasize to the lung.[138] The progression of carcinogen-induced (i.e., 3,2-dimethyl-4-amino biphenyl) prostate tumors has been inhibited by silymarin.[145] Finally, in the TRAMP model, apigenin,[146] EGCG,[147] genistein,[148] silibinin,[149] and xanthohumol[150] have been shown to inhibit progression of prostate tumorigenesis. However, the impact of the administered flavonoid may be dependent on its timing of administration with respect to the stage of tumorigenesis. For example, EGCG inhibited tumor progression when administered at an early, but not late, stage of tumorigenesis.[147] Exposure to genistein during late stage tumorigenesis accelerated its progression in an apparent estrogen- and PIK3CA-dependent manner.[151]

Epidemiological studies that have been performed to assess the impact of flavonoid intake on the risk of developing prostate cancer include several prospective cohort studies and a case control study. A ten-year prospective study of 43, 268 men residing in the United States was performed.[152] In this group, 3,974 prostate cancer cases were identified. Total daily intake of dietary flavonoids ranged from 99 mg/day to 505 mg/day with a mean intake of 270 mg/day. Greater flavonoid intake was positively associated with total prostate risk. A 17.3-year prospective study of 58,279 men in the Netherlands was also performed.[153] Here, 3,362 prostate cancers were identified and flavonoid intake was focused on two subgroups, catechins and flavonols. Median intake of total catechin was 53.8 mg/day and total flavonol was 25.7 mg/day. No association between dietary flavonoid intake and overall prostate cancer risk was observed. However, there was an inverse association between flavonoid intake and risk of advanced stage prostate cancer. In addition, a multiethnic prospective study of 82,483 men in Hawaii and Los Angeles was performed.[154] During this eight-year study, 4,404 prostate cancer cases were identified. Mean daily intake of total isoflavones was 118 mg. No association between intake of total (or any specific) isoflavone and prostate cancer risk was observed. Finally, a case control study consisting of 1,294 prostate cancer cases and 1,451 control cases was performed in Italy.[155] Mean daily intake of total flavonoids was 153.2 mg. No association between intake of either total or any subclass of flavonoid and prostate cancer risk was observed.

Flavonoids and prostate cancer chemoprevention. Clinical intervention studies that have investigated the impact of flavonoids on prostate cancer have primarily focused on green tea extracts containing high levels of EGCG. A double-blind placebo controlled study was performed to examine the effects of green tea catechins on the development of prostate cancer.[156] Patients involved in the study were diagnosed with high grade prostate intraepithelial neoplasia. The patients were administered a total of 600 mg of green tea catechins (taken as three capsules per day). The catechin preparations were composed of 51.8% EGCG and 75.7% total green tea catechins. After 12 months, nine of the 30 patients (30%) receiving the catechin capsules were diagnosed with prostate cancer compared to 1 of 30 patients (3.3%) receiving the placebo control.

Similarly, in a randomized, double-blind placebo controlled study, patients diagnosed with prostate cancer were randomly assigned to receive Polyphenon E® (a standardized formation of green tea extracts described in Chapter 1) or placebo for three to six weeks.[157] Four capsules, representing a total of 800 mg EGCG, were taken daily by the patients in the Polyphenon E® group. Average serum levels of EGCG of the Polyphenon E®-treated patients were 146.6 pmol/ml with lesser amounts of additional catechins also detected. Serum levels of oxidative DNA damage (8-hydroxy-2′-deoxyguanosine, 8 OHdG), IGF-1, and IGF binding protein 3 were reduced in the majority of the patients, but did not reach statistically significant differences when comparing the Polyphenon E® group versus that of the placebo controls. Analyses of prostate tissue biopsies similarly indicated that Polyphenon E treatment did not significantly alter the expression levels of markers of proliferation (as indicated by Ki-67 expression), apoptosis (as indicated by levels of cleaved caspase 3), or angiogenesis (as indicated by microvessel density). Adverse effects associated with the use of Polyphenon E® were limited to nausea, diarrhea, and headaches experienced by some of the patients. A phase II clinical trial using Polyphenon E® (400 mg EGCG/day) is currently ongoing and involves patients diagnosed with high grade prostatic intraepithelial neoplasia and/or atypical small acinar proliferation (Clinicaltrials.gov #NCT00596011). The results reported thus far indicate that after 12 months, four of 49 subjects in the Polyphenon E® treated group versus six of 48 subjects in the placebo control group had progressed to a diagnosis of prostate cancer.

An additional phase II clinical trial where either green or black tea was administered to patients diagnosed with clinically localized prostate adenocarcinoma was recently reported.[158] Ninety-three subjects drank six cups of either green tea, black tea, or water (daily) for 3–8 weeks prior to radical prostatectomy. Six cups of green tea were found to contain approximately 1,010 mg of tea polyphenols and 562 mg of EGCG whereas black tea contained approximately 80 mg of tea polyphenols and 28 mg of EGCG. Tissue biopsies taken prior to and at the end of the intervention

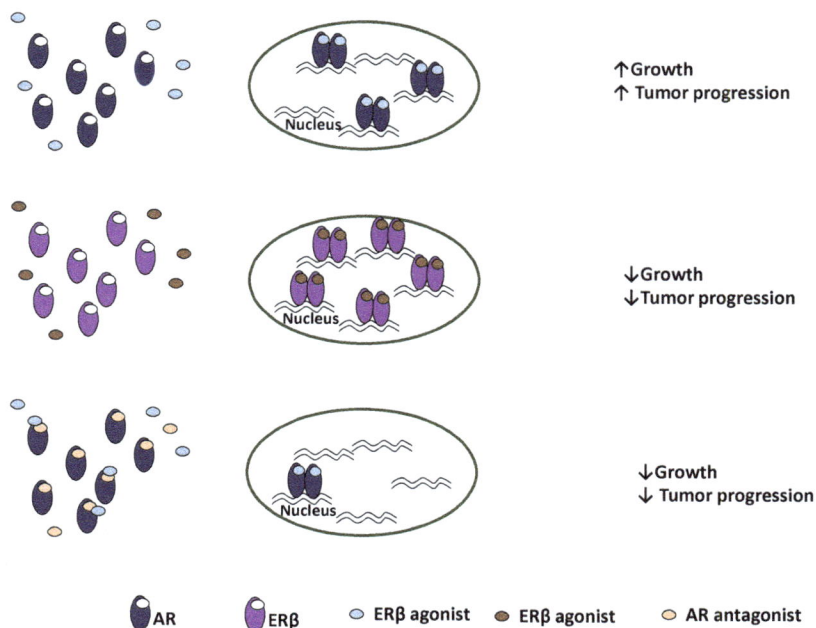

Figure 5.2. Putative impact of flavonoids in prostate tumor cells. The events that may be initiated by flavonoids with either estrogenic (i.e., selective activation of ERβ) or anti-androgenic actions are depicted. Endogenous androgens bind the AR which enhances the growth and progression of prostate tumor cells. Flavonoids such as liquiritigenin that interact with and activate ERβ may reduce tumor cell growth and inhibit tumor progression. Similarly, flavonoids with anti-androgenic properties such as 5-hydroxyflavone may competitively block endogenous androgen from binding the AR and thereby block its ability to impact tumor cell growth and progression.

revealed no significant differences between the treatment groups in the expression levels of markers of proliferation (Ki-67), apoptosis (i.e., BCL2 or BAX), or oxidative DNA damage (8 OHdG). A significant decrease in inflammation, as indicated by staining for nuclear accumulation of NFκB, was observed when comparing the results obtained in the green tea versus water control. Flavonoids, in particular, EGCG, could be detected in the prostate tissues from the patients who drank green tea, but not those that drank black tea or water.

Taken together, the results obtained from the clinical intervention trials indicate that there may be some promise associated with the use of green tea catechins to prevent the progression of prostate cancer. However, the evidence generated thus far is as yet inconclusive. Further, it is not yet clear whether or not the bioavailability and tissue accumulation of the catechins, in particular EGCG, is sufficient for exerting an effective chemopreventive effect.

References

1. Belfiore A, Perks CM. Grand challenges in cancer endocrinology: endocrine related cancers, an expanding concept. *Frontiers in Endocrinology.* 2013;4:141.
2. Hess RA. Estrogen in the adult male reproductive tract: A review. *Reprod Biol Endocrinol.* Jul 9, 2003;1:52.
3. Paterni I, Granchi C, Katzenellenbogen JA, Minutolo F. Estrogen receptors alpha (ERalpha) and beta (ERbeta): subtype-selective ligands and clinical potential. *Steroids.* Nov 2014;90:13–29.
4. Zhou W, Srinivasan S, Nawaz Z, Slingerland JM. ERalpha, SKP2 and E2F-1 form a feed forward loop driving late ERalpha targets and G1 cell cycle progression. *Oncogene.* May 1, 2014;33(18):2341–2353.
5. Knutson TP, Lange CA. Tracking progesterone receptor-mediated actions in breast cancer. *Pharmacology & Therapeutics.* Apr 2014;142(1):114–125.
6. Brisken C. Progesterone signalling in breast cancer: A neglected hormone coming into the limelight. *Nature Reviews. Cancer.* Jun 2013;13(6):385–396.
7. Matsumoto T, Sakari M, Okada M, *et al.* The androgen receptor in health and disease. *Annual Review of Physiology.* 2013;75:201–224.
8. McNamara KM, Sasano H. The intracrinology of breast cancer. *J Steroid Biochem Mol Biol.* Jan 2015;145C:172–178.
9. Zhou Y, Bolton EC, Jones JO. Androgens and androgen receptor signaling in prostate tumorigenesis. *Journal of Molecular Endocrinology.* Feb 2015;54(1):R15–R29.
10. Smith S, Sepkovic D, Bradlow HL, Auborn KJ. 3,3′-Diindolylmethane and genistein decrease the adverse effects of estrogen in LNCaP and PC-3 prostate cancer cells. *The Journal of Nutrition.* Dec 2008;138(12):2379–2385.
11. Gibson DA, Simitsidellis I, Collins F, Saunders PT. Evidence of androgen action in endometrial and ovarian cancers. *Endocrine-related Cancer.* Aug 2014;21(4):T203–218.

12. Collins-Burow BM, Burow ME, Duong BN, McLachlan JA. Estrogenic and antiestrogenic activities of flavonoid phytochemicals through estrogen receptor binding-dependent and -independent mechanisms. *Nutr Cancer.* 2000;38(2):229–244.

13. Lu DF, Yang LJ, Wang F, Zhang GL. Inhibitory effect of luteolin on estrogen biosynthesis in human ovarian granulosa cells by suppression of aromatase (CYP19). *J Agric Food Chem.* Aug 29, 2012;60(34):8411–8418.

14. Jiang Y, Gong P, Madak-Erdogan Z, et al. Mechanisms enforcing the estrogen receptor beta selectivity of botanical estrogens. *FASEB Journal: Official Publication of the Federation of American Societies for Experimental Biology.* Nov 2013;27(11):4406–4418.

15. Gong P, Madak-Erdogan Z, Li J, et al. Transcriptomic analysis identifies gene networks regulated by estrogen receptor alpha (ERalpha) and ERbeta that control distinct effects of different botanical estrogens. *Nuclear receptor signaling.* 2014;12:e001.

16. Sotoca AM, Ratman D, van der Saag P, et al. Phytoestrogen-mediated inhibition of proliferation of the human T47D breast cancer cells depends on the ERalpha/ERbeta ratio. *J Steroid Biochem Mol Biol.* Dec 2008;112(4–5):171–178.

17. Fang H, Tong W, Branham WS, et al. Study of 202 natural, synthetic, and environmental chemicals for binding to the androgen receptor. *Chemical Research in Toxicology.* Oct 2003;16(10):1338–1358.

18. Nishizaki Y, Ishimoto Y, Hotta Y, et al. Effect of flavonoids on androgen and glucocorticoid receptors based on *in vitro* reporter gene assay. *Bioorganic & Medicinal Chemistry Letters.* Aug 15, 2009;19(16):4706–4710.

19. Pihlajamaa P, Zhang FP, Saarinen L, Mikkonen L, Hautaniemi S, Janne OA. The phytoestrogen genistein is a tissue-specific androgen receptor modulator. *Endocrinology.* Nov 2011;152(11):4395–4405.

20. Ye L, Gho WM, Chan FL, Chen S, Leung LK. Dietary administration of the licorice flavonoid isoliquiritigenin deters the growth of MCF-7 cells overexpressing aromatase. *Int J Cancer.* Mar 1, 2009;124(5):1028–1036.

21. Du M, Yang X, Hartman JA, et al. Low-dose dietary genistein negates the therapeutic effect of tamoxifen in athymic nude mice. *Carcinogenesis.* Apr 2012;33(4):895–901.

22. Ju YH, Doerge DR, Woodling KA, Hartman JA, Kwak J, Helferich WG. Dietary genistein negates the inhibitory effect of letrozole on the growth of

aromatase-expressing estrogen-dependent human breast cancer cells (MCF-7Ca) *in vivo*. *Carcinogenesis*. Nov 2008;29(11):2162–2168.

23. Bracke ME, Depypere HT, Boterberg T, *et al*. Influence of tangeretin on tamoxifen's therapeutic benefit in mammary cancer. *Journal of the National Cancer Institute*. Feb 17, 1999;91(4):354–359.

24. Tu SH, Ho CT, Liu MF, *et al*. Luteolin sensitises drug-resistant human breast cancer cells to tamoxifen via the inhibition of cyclin E2 expression. *Food Chemistry*. Nov 15, 2013;141(2):1553–1561.

25. Long X, Fan M, Bigsby RM, Nephew KP. Apigenin inhibits antiestrogen-resistant breast cancer cell growth through estrogen receptor-alpha-dependent and estrogen receptor-alpha-independent mechanisms. *Molecular Cancer Therapeutics*. Jul 2008;7(7):2096–2108.

26. Li Y, Chen H, Hardy TM, Tollefsbol TO. Epigenetic regulation of multiple tumor-related genes leads to suppression of breast tumorigenesis by dietary genistein. *PLoS One*. 2013;8(1):e54369.

27. Curado MP. Breast cancer in the world: Incidence and mortality. *Salud publica de Mexico*. Sep–Oct 2011;53(5):372–384.

28. Ginsburg OM, Love RR. Breast cancer: A neglected disease for the majority of affected women worldwide. *The Breast Journal*. May–Jun 2011;17(3):289–295.

29. King MC, Levy–Lahad E, Lahad A. Population–based screening for BRCA1 and BRCA2: 2014 Lasker Award. *Jama*. Sep 17, 2014;312(11):1091–1092.

30. Anderson WF, Rosenberg PS, Prat A, Perou CM, Sherman ME. How many etiological subtypes of breast cancer: two, three, four, or more? *Journal of the National Cancer Institute*. Aug 2014;106(8).

31. Paul A, Paul S. The breast cancer susceptibility genes (BRCA) in breast and ovarian cancers. *Front Biosci (Landmark Ed)*. 2014;19:605–618.

32. Gutierrez C, Schiff R. HER2: biology, detection, and clinical implications. *Archives of Pathology & Laboratory Medicine*. Jan 2011;135(1):55–62.

33. Sestak I. Preventative therapies for healthy women at high risk of breast cancer. *Cancer Management and Research*. 2014;6:423–430.

34. Ahern TP, Lash TL, Damkier P, Christiansen PM, Cronin-Fenton DP. Statins and breast cancer prognosis: Evidence and opportunities. *The Lancet. Oncology*. Sep 2014;15(10):e461–468.

35. Takkouche B, Regueira-Mendez C, Etminan M. Breast cancer and use of nonsteroidal anti-inflammatory drugs: A meta-analysis. *Journal of the National Cancer Institute*. Oct 15, 2008;100(20):1439–1447.

36. Yang PM, Tseng HH, Peng CW, Chen WS, Chiu SJ. Dietary flavonoid fisetin targets caspase-3-deficient human breast cancer MCF-7 cells by induction of caspase-7-associated apoptosis and inhibition of autophagy. *Int J Oncol.* Feb 2012;40(2):469–478.

37. Li Y, Zhao H, Wang Y, *et al.* Isoliquiritigenin induces growth inhibition and apoptosis through downregulating arachidonic acid metabolic network and the deactivation of PI3K/Akt in human breast cancer. *Toxicol Appl Pharmacol.* Oct 1, 2013;272(1):37–48.

38. Park SH, Ham S, Kwon TH, *et al.* Luteolin induces cell cycle arrest and apoptosis through extrinsic and intrinsic signaling pathways in MCF-7 breast cancer cells. *Journal of Environmental Pathology, Toxicology and Oncology : Official Organ of the International Society for Environmental Toxicology and Cancer.* 2014;33(3):219–231.

39. Jin H, Lee WS, Eun SY, *et al.* Morin, a flavonoid from Moraceae, suppresses growth and invasion of the highly metastatic breast cancer cell line MDA-MB231 partly through suppression of the Akt pathway. *Int J Oncol.* Oct 2014;45(4):1629–1637.

40. Li H, Yang B, Huang J, *et al.* Naringin inhibits growth potential of human triple-negative breast cancer cells by targeting beta-catenin signaling pathway. *Toxicol Lett.* Jul 18, 2013;220(3):219–228.

41. Deng XH, Song HY, Zhou YF, Yuan GY, Zheng FJ. Effects of quercetin on the proliferation of breast cancer cells and expression of survivin. *Experimental and Therapeutic Medicine.* Nov 2013;6(5):1155–1158.

42. Tiwari P, Kumar A, Balakrishnan S, Kushwaha HS, Mishra KP. Silibinininduced apoptosis in MCF7 and T47D human breast carcinoma cells involves caspase-8 activation and mitochondrial pathway. *Cancer Investigation.* Jan 2011;29(1):12–20.

43. Hsu YL, Uen YH, Chen Y, Liang HL, Kuo PL. Tricetin, a dietary flavonoid, inhibits proliferation of human breast adenocarcinoma mcf-7 cells by blocking cell cycle progression and inducing apoptosis. *J Agric Food Chem.* Sep 23, 2009;57(18):8688–8695.

44. Shan Y, Cheng Y, Zhang Y, *et al.* Triticuside A, a dietary flavonoid, inhibits proliferation of human breast cancer cells via inducing apoptosis. *Nutr Cancer.* 2013;65(6):891–899.

45. Huang KF, Zhang GD, Huang YQ, Diao Y. Wogonin induces apoptosis and down-regulates survivin in human breast cancer MCF-7 cells by modulating

PI3K-AKT pathway. *International Immunopharmacology.* Feb 2012;12(2):334–341.

46. Delgado L, Fernandes I, Gonzalez-Manzano S, de Freitas V, Mateus N, Santos-Buelga C. Anti-proliferative effects of quercetin and catechin metabolites. *Food & Function.* Apr 2014;5(4):797–803.

47. Chen WY, Hsieh YA, Tsai CI, *et al.* Protoapigenone, a natural derivative of apigenin, induces mitogen-activated protein kinase-dependent apoptosis in human breast cancer cells associated with induction of oxidative stress and inhibition of glutathione S-transferase pi. *Investigational New Drugs.* Dec 2011;29(6):1347–1359.

48. Chen D, Landis-Piwowar KR, Chen MS, Dou QP. Inhibition of proteasome activity by the dietary flavonoid apigenin is associated with growth inhibition in cultured breast cancer cells and xenografts. *Breast Cancer Research: BCR.* 2007;9(6):R80.

49. Brusselmans K, Vrolix R, Verhoeven G, Swinnen JV. Induction of cancer cell apoptosis by flavonoids is associated with their ability to inhibit fatty acid synthase activity. *The Journal of Biological Chemistry.* Feb 18, 2005;280(7):5636–5645.

50. Maggiolini M, Bonofiglio D, Marsico S, *et al.* Estrogen receptor alpha mediates the proliferative but not the cytotoxic dose-dependent effects of two major phytoestrogens on human breast cancer cells. *Molecular Pharmacology.* Sep 2001;60(3):595–602.

51. Moon YJ, Shin BS, An G, Morris ME. Biochanin A inhibits breast cancer tumor growth in a murine xenograft model. *Pharm Res.* Sep 2008;25(9):2158–2163.

52. Mafuvadze B, Liang Y, Besch-Williford C, Zhang X, Hyder SM. Apigenin induces apoptosis and blocks growth of medroxyprogesterone acetate-dependent BT-474 xenograft tumors. *Hormones & Cancer.* Aug 2012;3(4):160–171.

53. Mafuvadze B, Cook M, Xu Z, Besch-Williford CL, Hyder SM. Effects of dietary apigenin on tumor latency, incidence and multiplicity in a medroxy-progesterone acetate-accelerated 7,12-dimethylbenz(a)anthracene-induced breast cancer model. *Nutr Cancer.* 2013;65(8):1184–1191.

54. Forghani P, Khorramizadeh MR, Waller EK. Silibinin inhibits accumulation of myeloid-derived suppressor cells and tumor growth of murine breast cancer. *Cancer Medicine.* Apr 2014;3(2):215–224.

55. Hsu YL, Wu LY, Hou MF, *et al.* Glabridin, an isoflavan from licorice root, inhibits migration, invasion and angiogenesis of MDA-MB-231 human breast adenocarcinoma cells by inhibiting focal adhesion kinase/Rho signaling pathway. *Mol Nutr Food Res.* Feb 2011;55(2):318–327.

56. Zheng H, Li Y, Wang Y, *et al.* Downregulation of COX-2 and CYP 4A signaling by isoliquiritigenin inhibits human breast cancer metastasis through preventing anoikis resistance, migration and invasion. *Toxicol Appl Pharmacol.* Oct 1, 2014;280(1):10–20.

57. Wang L, Ling Y, Chen Y, *et al.* Flavonoid baicalein suppresses adhesion, migration and invasion of MDA-MB-231 human breast cancer cells. *Cancer Letters.* Nov 1, 2010;297(1):42–48.

58. Lirdprapamongkol K, Sakurai H, Abdelhamed S, *et al.* A flavonoid chrysin suppresses hypoxic survival and metastatic growth of mouse breast cancer cells. *Oncol Rep.* Nov 2013;30(5):2357–2364.

59. Wang KL, Hsia SM, Chan CJ, *et al.* Inhibitory effects of isoliquiritigenin on the migration and invasion of human breast cancer cells. *Expert Opin Ther Targets.* Apr 2013;17(4):337–349.

60. Hui C, Qi X, Qianyong Z, Xiaoli P, Jundong Z, Mantian M. Flavonoids, flavonoid subclasses and breast cancer risk: A meta-analysis of epidemiologic studies. *PLoS One.* 2013;8(1):e54318.

61. Cho YA, Kim J, Park KS, *et al.* Effect of dietary soy intake on breast cancer risk according to menopause and hormone receptor status. *European Journal of Clinical Nutrition.* Sep 2010;64(9):924–932.

62. Wada K, Nakamura K, Tamai Y, *et al.* Soy isoflavone intake and breast cancer risk in Japan: From the Takayama study. *Int J Cancer.* Aug 15, 2013;133(4):952–960.

63. Zamora-Ros R, Ferrari P, Gonzalez CA, *et al.* Dietary flavonoid and lignan intake and breast cancer risk according to menopause and hormone receptor status in the European Prospective Investigation into Cancer and Nutrition (EPIC) Study. *Breast Cancer Research and Treatment.* May 2013;139(1):163–176.

64. Wang Y, Gapstur SM, Gaudet MM, Peterson JJ, Dwyer JT, McCullough ML. Evidence for an association of dietary flavonoid intake with breast cancer risk by estrogen receptor status is limited. *The Journal of Nutrition.* Oct 2014;144(10):1603–1611.

65. Luo J, Gao YT, Chow WH, *et al.* Urinary polyphenols and breast cancer risk: Results from the Shanghai Women's Health Study. *Breast Cancer Research and Treatment.* Apr 2010;120(3):693–702.

66. Morimoto Y, Maskarinec G, Park SY, *et al.* Dietary isoflavone intake is not statistically significantly associated with breast cancer risk in the Multiethnic Cohort. *The British Journal of Nutrition.* Sep 28, 2014;112(6):976–983.

67. Shike M, Doane AS, Russo L, *et al.* The effects of soy supplementation on gene expression in breast cancer: A randomized placebo-controlled study. *Journal of the National Cancer Institute.* Sep 2014;106(9).

68. Kang X, Zhang Q, Wang S, Huang X, Jin S. Effect of soy isoflavones on breast cancer recurrence and death for patients receiving adjuvant endocrine therapy. *CMAJ : Canadian Medical Association Journal = Journal de l'Association medicale canadienne.* Nov 23, 2010;182(17):1857–1862.

69. Jordan VC. Avoiding the bad and enhancing the good of soy supplements in breast cancer. *Journal of the National Cancer Institute.* Sep 2014;106(9).

70. Wu AH, Spicer D, Stanczyk FZ, Tseng CC, Yang CS, Pike MC. Effect of 2-month controlled green tea intervention on lipoprotein cholesterol, glucose, and hormone levels in healthy postmenopausal women. *Cancer Prev Res (Phila).* Mar 2012;5(3):393–402.

71. Crew KD, Brown P, Greenlee H, *et al.* Phase IB randomized, double-blinded, placebo-controlled, dose escalation study of polyphenon E in women with hormone receptor-negative breast cancer. *Cancer Prev Res (Phila).* Sep 2012;5(9):1144–1154.

72. Yu SS, Spicer DV, Hawes D, *et al.* Biological effects of green tea capsule supplementation in pre-surgery postmenopausal breast cancer patients. *Frontiers in Oncology.* 2013;3:298.

73. Vargas AN. Natural history of ovarian cancer. *Ecancermedicalscience.* 2014;8:465.

74. Rice LW. Hormone prevention strategies for breast, endometrial and ovarian cancers. *Gynecologic Oncology.* Aug 1, 2010;118(2):202–207.

75. Baandrup L, Kjaer SK, Olsen JH, Dehlendorff C, Friis S. Low-dose aspirin use and the risk of ovarian cancer in Denmark. *Annals of oncology : Official Journal of the European Society for Medical Oncology/ESMO.* Dec 23, 2014.

76. Trabert B, Ness RB, Lo-Ciganic WH, *et al.* Aspirin, nonaspirin nonsteroidal anti-inflammatory drug, and acetaminophen use and risk of invasive epithelial ovarian cancer: A pooled analysis in the Ovarian Cancer Association Consortium. *Journal of the National Cancer Institute.* Feb 2014;106(2):djt431.

77. Luo H, Jiang BH, King SM, Chen YC. Inhibition of cell growth and VEGF expression in ovarian cancer cells by flavonoids. *Nutr Cancer.* 2008; 60(6):800–809.

78. Luo H, Rankin GO, Li Z, Depriest L, Chen YC. Kaempferol induces apoptosis in ovarian cancer cells through activating p53 in the intrinsic pathway. *Food Chemistry.* Sep 15, 2011;128(2):513–519.

79. Chang HL, Su JH, Yeh YT, *et al.* Protoapigenone, a novel flavonoid, inhibits ovarian cancer cell growth *in vitro* and *in vivo. Cancer Letters.* Aug 18, 2008;267(1):85–95.

80. Cho HJ, Suh DS, Moon SH, *et al.* Silibinin inhibits tumor growth through downregulation of extracellular signal-regulated kinase and Akt *in vitro* and *in vivo* in human ovarian cancer cells. *J Agric Food Chem.* May 1, 2013;61(17):4089–4096.

81. Fan L, Ma Y, Liu Y, Zheng D, Huang G. Silymarin induces cell cycle arrest and apoptosis in ovarian cancer cells. *Eur J Pharmacol.* Nov 15, 2014;743: 79–88.

82. Zhou H, Lutterodt H, Cheng Z, Yu LL. Anti-Inflammatory and antiproliferative activities of trifolirhizin, a flavonoid from Sophora flavescens roots. *J Agric Food Chem.* Jun 10, 2009;57(11):4580–4585.

83. Hu XW, Meng D, Fang J. Apigenin inhibited migration and invasion of human ovarian cancer A2780 cells through focal adhesion kinase. *Carcinogenesis.* Dec 2008;29(12):2369–2376.

84. Fang J, Xia C, Cao Z, Zheng JZ, Reed E, Jiang BH. Apigenin inhibits VEGF and HIF-1 expression via PI3K/AKT/p70S6K1 and HDM2/p53 pathways. *FASEB Journal: Official Publication of the Federation of American Societies for Experimental Biology.* Mar 2005;19(3):342–353.

85. He J, Xu Q, Wang M, *et al.* Oral Administration of Apigenin Inhibits Metastasis through AKT/P70S6K1/MMP-9 Pathway in Orthotopic Ovarian Tumor Model. *International Journal of Molecular Sciences.* 2012;13(6):7271–7282.

86. Li N, Sun C, Zhou B, *et al.* Low concentration of quercetin antagonizes the cytotoxic effects of anti-neoplastic drugs in ovarian cancer. *PLoS One.* 2014;9(7):e100314.

87. Wang L, Lee IM, Zhang SM, Blumberg JB, Buring JE, Sesso HD. Dietary intake of selected flavonols, flavones, and flavonoid-rich foods and risk of cancer in middle-aged and older women. *The American Journal of Clinical Nutrition.* Mar 2009;89(3):905–912.

88. Gates MA, Vitonis AF, Tworoger SS, *et al.* Flavonoid intake and ovarian cancer risk in a population-based case-control study. *Int J Cancer.* Apr 15, 2009;124(8):1918–1925.

89. Neill AS, Ibiebele TI, Lahmann PH, *et al*. Dietary phyto-oestrogens and the risk of ovarian and endometrial cancers: findings from two Australian case-control studies. *The British Journal of Nutrition*. Apr 28, 2014;111(8):1430–1440.

90. Trudel D, Labbe DP, Araya-Farias M, *et al*. A two-stage, single-arm, phase II study of EGCG-enriched green tea drink as a maintenance therapy in women with advanced stage ovarian cancer. *Gynecologic Oncology*. Nov 2013;131(2):357–361.

91. Cramer DW. The epidemiology of endometrial and ovarian cancer. *Hematology/Oncology Clinics of North America*. Feb 2012;26(1):1–12.

92. Murali R, Soslow RA, Weigelt B. Classification of endometrial carcinoma: More than two types. *The Lancet. Oncology*. Jun 2014;15(7):e268–278.

93. Friel AM, Growdon WB, McCann CK, *et al*. Mouse models of uterine corpus tumors: Clinical significance and utility. *Frontiers in Bioscience*. 2010;2:882–905.

94. Cheng H, Liu P, Zhang F, *et al*. A genetic mouse model of invasive endometrial cancer driven by concurrent loss of Pten and Lkb1 Is highly responsive to mTOR inhibition. *Cancer Res*. Jan 1, 2014;74(1):15–23.

95. Lu KH, Loose DS, Yates MS, *et al*. Prospective multicenter randomized intermediate biomarker study of oral contraceptive versus depo-provera for prevention of endometrial cancer in women with Lynch syndrome. *Cancer Prev Res (Phila)*. Aug 2013;6(8):774–781.

96. Brasky TM, Moysich KB, Cohn DE, White E. Non-steroidal anti-inflammatory drugs and endometrial cancer risk in the VITamins And Lifestyle (VITAL) cohort. *Gynecologic Oncology*. Jan 2013;128(1):113–119.

97. Schuler KM, Rambally BS, DiFurio MJ, *et al*. Antiproliferative and metabolic effects of metformin in a preoperative window clinical trial for endometrial cancer. *Cancer Medicine*. Feb 2015;4(2):161–173.

98. Tabrizi AD, Melli MS, Foroughi M, Ghojazadeh M, Bidadi S. Antiproliferative effect of metformin on the endometrium — a clinical trial. *Asian Pacific Journal of Cancer Prevention: APJCP*. 2014;15(23):10067–10070.

99. Manohar M, Fatima I, Saxena R, Chandra V, Sankhwar PL, Dwivedi A. (–)-Epigallocatechin-3-gallate induces apoptosis in human endometrial adenocarcinoma cells via ROS generation and p38 MAP kinase activation. *J Nutr Biochem*. Jun 2013;24(6):940–947.

100. Cho JH, Lee JG, Yang YI, *et al*. Eupatilin, a dietary flavonoid, induces G2/M cell cycle arrest in human endometrial cancer cells. *Food Chem Toxicol*. Aug 2011;49(8):1737–1744.

101. Tong JS, Zhang QH, Huang X, *et al.* Icaritin causes sustained ERK1/2 activation and induces apoptosis in human endometrial cancer cells. *PLoS One.* 2011;6(3):e16781.

102. Sampey BP, Lewis TD, Barbier CS, Makowski L, Kaufman DG. Genistein effects on stromal cells determines epithelial proliferation in endometrial co-cultures. *Experimental and Molecular Pathology.* Jun 2011;90(3): 257–263.

103. Gaete L, Tchernitchin AN, Bustamante R, *et al.* Daidzein-estrogen interaction in the rat uterus and its effect on human breast cancer cell growth. *Journal of Medicinal Food.* Dec 2012;15(12):1081–1090.

104. Gaete L, Tchernitchin AN, Bustamante R, *et al.* Genistein selectively inhibits estrogen-induced cell proliferation and other responses to hormone stimulation in the prepubertal rat uterus. *Journal of Medicinal Food.* Dec 2011;14(12):1597–1603.

105. Kakehashi A, Tago Y, Yoshida M, *et al.* Hormonally active doses of isoflavone aglycones promote mammary and endometrial carcinogenesis and alter the molecular tumor environment in Donryu rats. *Toxicological sciences: An Official Journal of the Society of Toxicology.* Mar 2012;126(1):39–51.

106. Rossi M, Edefonti V, Parpinel M, *et al.* Proanthocyanidins and other flavonoids in relation to endometrial cancer risk: A case-control study in Italy. *British Journal of Cancer.* Oct 1, 2013;109(7):1914–1920.

107. Quaas AM, Kono N, Mack WJ, *et al.* Effect of isoflavone soy protein supplementation on endometrial thickness, hyperplasia, and endometrial cancer risk in postmenopausal women: A randomized controlled trial. *Menopause.* Aug 2013;20(8):840–844.

108. Ollberding NJ, Lim U, Wilkens LR, *et al.* Legume, soy, tofu, and isoflavone intake and endometrial cancer risk in postmenopausal women in the multiethnic cohort study. *Journal of the National Cancer Institute.* Jan 4, 2012;104(1):67–76.

109. Yeh CR, Da J, Song W, Fazili A, Yeh S. Estrogen receptors in prostate development and cancer. *American Journal of Clinical and Experimental Urology.* 2014;2(2):161–168.

110. Grabowska MM, DeGraff DJ, Yu X, *et al.* Mouse models of prostate cancer: Picking the best model for the question. *Cancer Metastasis Reviews.* Sep 2014;33(2–3):377–397.

111. Strope SA, Andriole GL. Update on chemoprevention for prostate cancer. *Current Opinion in Urology.* May 2010;20(3):194–197.

112. Lacy JM, Kyprianou N. A tale of two trials: The impact of 5alpha-reductase inhibition on prostate cancer (Review). *Oncology Letters.* Oct 2014;8(4): 1391–1396.

113. Thompson IM, Jr., Goodman PJ, Tangen CM, *et al.* Long-term survival of participants in the prostate cancer prevention trial. *The New England Journal of Medicine.* Aug 15, 2013;369(7):603–610.

114. Lin PH, Aronson W, Freedland SJ. Nutrition, dietary interventions and prostate cancer: The latest evidence. *BMC Medicine.* 2015;13:3.

115. Shen KH, Hung SH, Yin LT, *et al.* Acacetin, a flavonoid, inhibits the invasion and migration of human prostate cancer DU145 cells via inactivation of the p38 MAPK signaling pathway. *Mol Cell Biochem.* Jan 2010;333(1–2):279–291.

116. Ni F, Gong Y, Li L, Abdolmaleky HM, Zhou JR. Flavonoid ampelopsin inhibits the growth and metastasis of prostate cancer *in vitro* and in mice. *PLoS One.* 2012;7(6):e38802.

117. Kaur P, Shukla S, Gupta S. Plant flavonoid apigenin inactivates Akt to trigger apoptosis in human prostate cancer: An *in vitro* and *in vivo* study. *Carcinogenesis.* Nov 2008;29(11):2210–2217.

118. Shukla S, MacLennan GT, Flask CA, *et al.* Blockade of beta-catenin signaling by plant flavonoid apigenin suppresses prostate carcinogenesis in TRAMP mice. *Cancer Res.* Jul 15, 2007;67(14):6925–6935.

119. Miocinovic R, McCabe NP, Keck RW, Jankun J, Hampton JA, Selman SH. *In vivo* and *in vitro* effect of baicalein on human prostate cancer cells. *Int J Oncol.* Jan 2005;26(1):241–246.

120. Sartor L, Pezzato E, Dona M, *et al.* Prostate carcinoma and green tea: (−) epigallocatechin-3-gallate inhibits inflammation-triggered MMP-2 activation and invasion in murine TRAMP model. *Int J Cancer.* Dec 10, 2004;112(5):823–829.

121. Suh Y, Afaq F, Khan N, Johnson JJ, Khusro FH, Mukhtar H. Fisetin induces autophagic cell death through suppression of mTOR signaling pathway in prostate cancer cells. *Carcinogenesis.* Aug 2010;31(8):1424–1433.

122. Khan N, Afaq F, Syed DN, Mukhtar H. Fisetin, a novel dietary flavonoid, causes apoptosis and cell cycle arrest in human prostate cancer LNCaP cells. *Carcinogenesis.* May 2008;29(5):1049–1056.

123. Lee CJ, Wilson L, Jordan MA, Nguyen V, Tang J, Smiyun G. Hesperidin suppressed proliferations of both human breast cancer and androgen-dependent prostate cancer cells. *Phytother Res.* Jan 2010;24 Suppl 1:S15–19.

124. Lee YM, Lim do Y, Choi HJ, Jung JI, Chung WY, Park JH. Induction of cell cycle arrest in prostate cancer cells by the dietary compound isoliquiritigenin. *Journal of Medicinal Food.* Feb 2009;12(1):8–14.

125. Jung JI, Lim SS, Choi HJ, *et al.* Isoliquiritigenin induces apoptosis by depolarizing mitochondrial membranes in prostate cancer cells. *J Nutr Biochem.* Oct 2006;17(10):689–696.

126. Pratheeshkumar P, Son YO, Budhraja A, *et al.* Luteolin inhibits human prostate tumor growth by suppressing vascular endothelial growth factor receptor 2-mediated angiogenesis. *PLoS One.* 2012;7(12):e52279.

127. Lee J, Lee S, Kim SL, *et al.* Corn silk maysin induces apoptotic cell death in PC-3 prostate cancer cells via mitochondria-dependent pathway. *Life Sciences.* Dec 5, 2014;119(1–2):47–55.

128. Chen J, Creed A, Chen AY, *et al.* Nobiletin suppresses cell viability through AKT pathways in PC-3 and DU-145 prostate cancer cells. *BMC Pharmacology & Toxicology.* 2014;15:59.

129. Vijayababu MR, Kanagaraj P, Arunkumar A, Ilangovan R, Dharmarajan A, Arunakaran J. Quercetin induces p53-independent apoptosis in human prostate cancer cells by modulating Bcl-2-related proteins: a possible mediation by IGFBP-3. *Oncology Research.* 2006;16(2):67–74.

130. Nagaprashantha LD, Vatsyayan R, Singhal J, *et al.* Anti-cancer effects of novel flavonoid vicenin-2 as a single agent and in synergistic combination with docetaxel in prostate cancer. *Biochem Pharmacol.* Nov 1, 2011;82(9):1100–1109.

131. Shukla S, Gupta S. Apigenin-induced cell cycle arrest is mediated by modulation of MAPK, PI3K-Akt, and loss of cyclin D1 associated retinoblastoma dephosphorylation in human prostate cancer cells. *Cell Cycle.* May 2, 2007;6(9):1102–1114.

132. Chiu FL, Lin JK. Downregulation of androgen receptor expression by luteolin causes inhibition of cell proliferation and induction of apoptosis in human prostate cancer cells and xenografts. *The Prostate.* Jan 1, 2008;68(1):61–71.

133. Mahmoud AM, Zhu T, Parray A, *et al.* Differential effects of genistein on prostate cancer cells depend on mutational status of the androgen receptor. *PLoS One.* 2013;8(10):e78479.

134. Tang SN, Singh C, Nall D, Meeker D, Shankar S, Srivastava RK. The dietary bioflavonoid quercetin synergizes with epigallocathechin gallate (EGCG) to inhibit prostate cancer stem cell characteristics, invasion, migration and epithelial-mesenchymal transition. *Journal of Molecular Signaling.* 2010;5:14.

135. Xu R, Zhang Y, Ye X, *et al.* Inhibition effects and induction of apoptosis of flavonoids on the prostate cancer cell line PC-3 *in vitro. Food Chemistry.* May 1, 2013;138(1):48–53.

136. Singh RP, Agrawal P, Yim D, Agarwal C, Agarwal R. Acacetin inhibits cell growth and cell cycle progression, and induces apoptosis in human prostate cancer cells: Structure-activity relationship with linarin and linarin acetate. *Carcinogenesis.* Apr 2005;26(4):845–854.

137. Ko CC, Chen YJ, Chen CT, *et al.* Chemical proteomics identifies heterogeneous nuclear ribonucleoprotein (hnRNP) A1 as the molecular target of quercetin in its anti-cancer effects in PC-3 cells. *The Journal of biological Chemistry.* Aug 8, 2014;289(32):22078–22089.

138. Zhou Q, Yan B, Hu X, Li XB, Zhang J, Fang J. Luteolin inhibits invasion of prostate cancer PC3 cells through E-cadherin. *Molecular Cancer Therapeutics.* Jun 2009;8(6):1684–1691.

139. Bhat FA, Sharmila G, Balakrishnan S, *et al.* Quercetin reverses EGF-induced epithelial to mesenchymal transition and invasiveness in prostate cancer (PC-3) cell line via EGFR/PI3K/Akt pathway. *J Nutr Biochem.* Nov 2014;25(11):1132–1139.

140. Gao K, Henning SM, Niu Y, *et al.* The citrus flavonoid naringenin stimulates DNA repair in prostate cancer cells. *J Nutr Biochem.* Feb 2006;17(2):89–95.

141. Khan N, Asim M, Afaq F, Abu Zaid M, Mukhtar H. A novel dietary flavonoid fisetin inhibits androgen receptor signaling and tumor growth in athymic nude mice. *Cancer Res.* Oct 15, 2008;68(20):8555–8563.

142. Slusarz A, Jackson GA, Day JK, *et al.* Aggressive prostate cancer is prevented in ERalphaKO mice and stimulated in ERbetaKO TRAMP mice. *Endocrinology.* Sep 2012;153(9):4160–4170.

143. Chang HL, Wu YC, Su JH, Yeh YT, Yuan SS. Protoapigenone, a novel flavonoid, induces apoptosis in human prostate cancer cells through activation of p38 mitogen-activated protein kinase and c-Jun NH2-terminal kinase 1/2. *The Journal of Pharmacology and Experimental Therapeutics.* Jun 2008;325(3):841–849.

144. Singh RP, Dhanalakshmi S, Tyagi AK, Chan DC, Agarwal C, Agarwal R. Dietary feeding of silibinin inhibits advance human prostate carcinoma growth in athymic nude mice and increases plasma insulin-like growth factor-binding protein-3 levels. *Cancer Res.* Jun 1, 2002;62(11):3063–3069.

145. Kohno H, Suzuki R, Sugie S, Tsuda H, Tanaka T. Dietary supplementation with silymarin inhibits 3,2'-dimethyl-4-aminobiphenyl-induced prostate carcinogenesis in male F344 rats. *Clinical Cancer Research: An Official Journal of the American Association for Cancer Research.* Jul 1, 2005;11(13):4962–4967.

146. Shukla S, MacLennan GT, Fu P, Gupta S. Apigenin attenuates insulin-like growth factor-I signaling in an autochthonous mouse prostate cancer model. *Pharm Res.* Jun 2012;29(6):1506–1517.

147. Harper CE, Patel BB, Wang J, Eltoum IA, Lamartiniere CA. Epigallocatechin-3-Gallate suppresses early stage, but not late stage prostate cancer in TRAMP mice: Mechanisms of action. *The Prostate.* Oct 1, 2007;67(14): 1576–1589.

148. El Touny LH, Banerjee PP. Genistein induces the metastasis suppressor kangai-1 which mediates its anti-invasive effects in TRAMP cancer cells. *Biochemical and Biophysical Research Communications.* Sep 14, 2007;361(1):169–175.

149. Singh RP, Raina K, Sharma G, Agarwal R. Silibinin inhibits established prostate tumor growth, progression, invasion, and metastasis and suppresses tumor angiogenesis and epithelial-mesenchymal transition in transgenic adenocarcinoma of the mouse prostate model mice. *Clinical Cancer Research: An Official Journal of the American Association for Cancer Research.* Dec 1, 2008;14(23):7773–7780.

150. Vene R, Benelli R, Minghelli S, Astigiano S, Tosetti F, Ferrari N. Xanthohumol impairs human prostate cancer cell growth and invasion and diminishes the incidence and progression of advanced tumors in TRAMP mice. *Molecular Medicine.* 2012;18:1292–1302.

151. El Touny LH, Banerjee PP. Identification of a biphasic role for genistein in the regulation of prostate cancer growth and metastasis. *Cancer Res.* Apr 15, 2009;69(8):3695–3703.

152. Wang Y, Stevens VL, Shah R, *et al.* Dietary flavonoid and proanthocyanidin intakes and prostate cancer risk in a prospective cohort of US men. *American Journal of Epidemiology.* Apr 15, 2014;179(8):974–986.

153. Mursu J, Nurmi T, Tuomainen TP, Salonen JT, Pukkala E, Voutilainen S. Intake of flavonoids and risk of cancer in Finnish men: The Kuopio Ischaemic Heart Disease Risk Factor Study. *Int J Cancer.* Aug 1, 2008;123(3):660–663.

154. Park SY, Murphy SP, Wilkens LR, Henderson BE, Kolonel LN, Multiethnic Cohort S. Legume and isoflavone intake and prostate cancer risk: The Multiethnic Cohort Study. *Int J Cancer.* Aug 15, 2008;123(4):927–932.

155. Bosetti C, Bravi F, Talamini R, *et al.* Flavonoids and prostate cancer risk: A study in Italy. *Nutr Cancer.* 2006;56(2):123–127.

156. Bettuzzi S, Brausi M, Rizzi F, Castagnetti G, Peracchia G, Corti A. Chemoprevention of human prostate cancer by oral administration of green tea catechins in volunteers with high-grade prostate intraepithelial neoplasia: A preliminary report from a one-year proof-of-principle study. *Cancer Res.* Jan 15, 2006;66(2):1234–1240.

157. Nguyen MM, Ahmann FR, Nagle RB, *et al.* Randomized, double-blind, placebo-controlled trial of polyphenon E in prostate cancer patients before prostatectomy: Evaluation of potential chemopreventive activities. *Cancer Prev Res (Phila).* Feb 2012;5(2):290–298.

158. Henning SM, Wang P, Said JW, *et al.* Randomized clinical trial of brewed green and black tea in men with prostate cancer prior to prostatectomy. *The Prostate.* Apr 2015;75(5):550–559.

Summary and Future Directions* $\quad$ **6**

For decades, we have worked toward the goal of chemoprevention: to reduce the burden of cancer by using dietary or pharmacological substances capable of inhibiting tumor formation and progression. The development of these agents has been fraught with both successes and failures when ultimately tested in human subjects.[1] The most commonly observed failures have included unexpected tumor-promoting effects, significant adverse effects, or lack of beneficial, anti-tumor effects. However, it should be recognized that successes have also been achieved, as indicated by the 13 pharmacological agents that have been approved by the U.S. Food and Drug Administration for their use as chemopreventive agents. These agents include vaccines to prevent human papilloma viral (HPV) infections and reduce the risk of cervical, vaginal, and anal cancers, anti-estrogens (tamoxifen and raloxifene) to treat women who are at high risk of developing breast cancer, and the COX-2 inhibiting, anti-inflammatory agent, Celecoxib, to treat individuals who are at high risk for developing colorectal cancer.

The intense interest in flavonoids as anti-inflammatory and chemopreventive agents is indicated by the voluminous scientific work published in this area and the increasing number of clinical trials being performed using flavonoids or flavonoid-containing substances as clinical interventions. This is apparent by the extensive literature that has largely focused on the use of individual flavonoids to inhibit cancer-relevant events as discussed in the preceding chapters. Further, a recent PubMed search

---

*****Abbreviations:** AKT, v-akt murine thymoma viral oncogene homolog 1; CA125, cancer antigen 125; EGCG, (-)-epigallocatechin-3- gallate; He4, human epididymis protein 4; IL8, interleukin 8; NFκB, nuclear factor of kappa light polypeptide gene enhancer in B cells 1; PIK3CA; phosphatidylinositol-4,5-bisphosphate 3-kinase, catalytic subunit alpha; PSA, prostate-specific antigen; mTOR, mechanistic target of rapamycin.

using the terms "flavonoid" and "cancer" resulted in the recovery of over 13,000 published manuscripts, and use of the terms "flavonoid" and "anti-inflammation" yielded more than 8,000 publications. A search of the ClinicalTrials.gov database (accessed on April 2, 2015), revealed that approximately 32 clinical trials are currently investigating the effects of green tea extracts or other supplements on cancer-related endpoints. Additional ongoing trials that involve patients diagnosed with a variety of cancers are also being performed where sources of flavonoids used include berry extracts, cocoa powder, and broccoli sprouts, as well as individual flavonoids such as quercetin and genistein. A high number of ongoing clinical trials also include testing the impact of flavonoids or flavonoid-containing substances on endpoints related to other disease states including diabetes and cardiovascular diseases.

The popularity of flavonoids as potential therapeutic agents is due primarily to the low cost associated with their preparation and their excellent safety profiles. In the clinical intervention trials previously discussed, the most commonly observed adverse effects were relatively mild and involved primarily gastrointestinal complaints even when high concentrations of flavonoids were administered. However, use of flavonoids as adjunct therapies to be co-administered with a commonly used chemotherapeutic drug in an attempt to enhance tumoricidal activities is also gaining traction. This proposed clinical use should be viewed with caution as the adverse effects that may occur as a consequence of the co-administration of these two substances are difficult to predict. As mentioned in Chapter 2, many flavonoids are known to inhibit drug metabolism and drug export. Thus, the co-administration of such a flavonoid with a chemotherapeutic drug which typically harbors a narrow therapeutic index could potentially result in enhanced and significant adverse effects in human subjects. The severe abdominal pain experienced by patients with pancreatic cancer who were co-treated with curcumin and gemcitabine is an example of this type of supplement-drug interaction which may have arisen from enhanced toxicity of gemcitabine.[2] This issue should be further explored using pharmacokinetic studies.

As can be observed from our extensive discussion, the question of whether a high dietary intake of flavonoids can be effective in preventing the onset or progression of cancer remains unresolved. This fact remains despite the plethora of data generated from decades of laboratory-based studies using either cultured cells or animal models of human cancers or from epidemiological studies representing many different forms of cancer occurring in a large swath of the global population. Factors thought to contribute to the uncertainty surrounding the chemopreventive status of flavonoids include (1) the complexity of cancer as it exists as a disease state within the human population, (2) the lack of a definitive mechanism by which flavonoids exert their beneficial effects, (3) uncertainties regarding the bioavailability of the pharmacologically active chemical form, (4) the undefined role of the microbiome, and (5) the lack of appropriate biomarkers.

Progress on understanding the chemopreventive actions of flavonoids is hindered by the complexity of cancer as a disease state. Using an ever-expanding set of molecular and genetic tools over the past several years, we have begun to unravel the complexity of cancer as a disease state.[3] Even within a single tissue type, cancer is not a single disease, but represents a multitude of heterogenous diseases, each with its unique pathologies, underlying genetic makeup, potential for metastasis, and response to treatment. It is now realized that a tumor obtained from a patient diagnosed with a metastatic disease is highly likely to harbor a unique set of genetic aberrations.

The complexity of cancer may be best illustrated by our current knowledge of breast cancer. Breast tumors display significant variability depending on whether they have developed in pre-menopausal versus post-menopausal women, their molecular signatures, their overall prognosis, and whether or not the tumors will respond to endocrine-based therapies.[4,5] Breast cancer is now known to be composed of several different subtypes, only some of which are driven by aberrant estrogen receptor (i.e., estrogen receptor α) signaling. The roles of other steroid receptors, in particular estrogen receptor β, progesterone receptor, and androgen receptor, as well as signaling pathways such as PIK3CA/AKT/mTOR in human breast cancers, have only recently come into focus. Our attempts to identify the originating, progenitor tumor cells have led to our realization that a hierarchy of cells exists within the mammary epithelium. The pluripotent stem cells are thought to be the origin of estrogen receptor negative breast tumor cells which are amongst the most aggressive and difficult to treat. Stem cells that can easily switch from self-renewal and differentiation and expand during chronic inflammation are highly likely to become tumor cells. Further, the microenvironment of the tumor cells often provides a condition of optimal growth for the tumor cell, but may also be a target for agents capable of "normalizing" the environment such that the growth of the aberrant cell is kept in check. Our attempts to target the tumor cell's neighbors, rather than the tumor cell itself, have been met with some failures. For example, targeting the endothelial cells within the tumor microenvironment was initially met with considerable enthusiasm. However, when ultimately tested in clinical trials, drugs that targeted the key regulators of angiogenesis failed to demonstrate a beneficial effect and caused severe side effects. It remains to be seen whether other cell types within the tumor microenvironment, in particular, the immune cells, will serve as more effective targets for controlling tumor growth.

With respect to breast cancer chemoprevention, the administration of either tamoxifen or raloxifene is recommended for women considered to be at high risk of developing breast cancer.[6] However, it has been difficult to clearly identify the women most likely to benefit from these pharmacological interventions and their use can be accompanied by significant side effects. While chemoprevention using flavonoids either via increased dietary consumption or via the use of supplements holds promise due to their perceived lack of serious side effects, whether or not substantial benefits can be achieved is yet to be determined. This issue is perhaps best illustrated by the ongoing debate pertaining to the intake of isoflavones in pre-versus post-menopausal women.[7] However, as we gain a better understanding of the originating tumor cell types, its etiology, defining molecular attributes, and attributes specific to the host, we can make progress in determining which individuals are most likely to benefit from the chemopreventive actions of flavonoids.

Progress in understanding the mechanisms by which flavonoids may exert their beneficial effects. The mechanisms by which flavonoids exert their chemopreventive effects appear to involve their modulation of a circuitry of all six of the hallmarks of cancer as well as inhibiting oxidative stress and inflammation. The lack of a single, well-defined mechanism of action is a major hindrance for designing informative human epidemiological studies due to difficulties in ascertaining the appropriate biomarker(s) to be used for evaluating efficacy of treatment. However, flavonoids, like aspirin, may exert their chemopreventive effects due to a composite of events that all converge on inhibition of inflammation and perhaps, signaling of the transcription factor NFκB. Studies performed using animal models of experimental colitis (discussed in Chapter 3) indicate that many flavonoids can act as good anti-inflammatory agents, particularly when their bioavailability is improved. The anti-inflammatory properties of flavonoids also appear to be enhanced when administered as a mixture, such as lyophilized berry extracts or green tea extracts, as compared to the effectiveness of isolated, individual compounds. In contrast to the anti-inflammatory properties of flavonoids, their impact on either estrogen or androgen signaling has been more difficult to predict. In many cases, flavonoids appear to exhibit partial agonist activities with respect to their interactions with estrogen or androgen receptors. This may reflect the complexity of hormone signaling within steroid hormone-responsive tissues and the ability of many chemicals to act as either selective estrogen receptor modulators (SERMS) or selective androgen receptor modulators (SARMS) in cell context-dependent manners.

While the approaches used in the laboratory have substantially improved in their ability to mimic biological events that occur during human tumorigenesis, they are still inadequate in mirroring the complexity of the human disease state. The majority of studies discussed in previous chapters have examined the impact of flavonoids on various aspects of the tumorigenic process using cultured human tumor cells that may not appropriately represent the genetic complexity of tumors that occur within the human population. The impact of an individual flavonoid, in its parent form and often administered at relatively high concentrations, on signaling systems within a monolayer of a single cell type likely does not appropriately represent events that are initiated by an ingested flavonoid on the signaling network within and amongst the many types of cells that reside within the tumor microenvironment. Improvements that may be made in this regard include the use of tissue engineering platforms that allow for the integration of subcellular networking within the tumor microenvironment.[8] Further, three-dimensional culture systems that better mimic the cellular matrix of the *in vivo* tissue can be used. For example, a three-dimensional breast tissue culture model that is responsive to estrogen-induced cell proliferation has recently been developed.[9] Finally, the impact of tumor genetic diversity on the effects of flavonoids may be assessed using the circulating tumor cells that have been obtained from the patient and cultured.[10]

Another issue compromising our ability to accurately predict events that occur within human cancers is the inadequacy of the animal models used to study the disease state. The chemopreventive effects of individual flavonoids have been studied *in vivo* using immune compromised mice bearing xenografts of human tumor cell lines, using carcinogen-induced tumors in rats and mice and using transgenic mouse models. Of these, the least clinically predictive model is that using tumor xenografts.[11] This model can be improved by transplanting freshly excised tumors obtained from individual patients into the immune compromised mouse. However, a major disadvantage of using any tumor explant model is that it tests the impact of potential chemopreventive agents on fully formed tumor cells. These tumor cells represent the ultimate stage of tumorigenesis as they have developed a plethora of survival tactics. Further, the growth of the explanted tumor cell is being conducted outside of its native tumor microenvironment and thus the potential roles of its constitutive neighbors, such as the immune cells that have infiltrated the original tumor mass, remains unexamined. A better strategy may be to introduce the chemopreventive agent before a tumor cell has built its nearly impenetrable defenses. This latter scenario is recapitulated in the carcinogen-induced and transgenic mouse models of carcinogenesis and hence are preferred approaches to be used for testing chemopreventive agents. Clinical relevance is reportedly highest when the results have been obtained using the carcinogen-induced model as it is thought to represent the multi-step events that occur during tumorigenesis and in the presence of an intact immune system. The clinical predictiveness of transgenic models continues to improve as new strains are developed that more accurately portray the human tumorigenic process in the tissues of interest. The idea that chemopreventive agents may work best during the initial stages of carcinogenesis is supported by several of the studies reporting that the impact of the administered flavonoid was dependent on its timing of its administration.[12,13] If these findings translate into events that occur within the human population, it implies that a "window of responsiveness" may exist representing the stage of growth and/or progression that an individual's tumor is most responsive to the flavonoid's chemopreventive effects. Efforts to identify this "window of responsiveness" as well as individuals that would most benefit from the effects of flavonoids may be aided by enhancing our understanding of the relationship between flavonoid intake and putative biomarkers such as IL8 and perhaps other pro-inflammatory markers.[14,15]

Progress on understanding and improving the bioavailability of flavonoids. An active area of research is focused on understanding the bioavailability of flavonoids, their pharmacokinetics, and the impact of their structural variations. Quercetin and tea catechins are among the best characterized flavonoids, with respect to their bioavailability and metabolism in human subjects.[16,17] Analyses of quercetin metabolism revealed that one hour following the consumption of a quercetin-rich food source (onion), 23 different metabolic species of quercetin could be identified in the plasma and urine.[16,18] The three major metabolites formed were 3'methylquercetin-3-*O*-β-D-glucuronide, quercetin-3-*O*-β-D-glucuronide, and quercetin-3'-*O*-sulfate. Interestingly, 3'-methylquercetin-3-*O*-β-D-glucuronide has been described as exhibiting relatively high potency with respect to antioxidant activities and other beneficial effects. Similar studies have examined the bioavailability of flavonoids in human subjects after their consumption of green tea.[17,19] Consumption of two cups of green tea was found to result in peak plasma levels of 0.15 μM EGCG. Further, following tea consumption, ten different catechin metabolites, consisting of O-methylated, sulfated, and glucuronide conjugates were present in the plasma. In the urine, 15 similarly conjugated metabolites were formed and accounted for 8% of the total intake. With respect to tissue concentrations, four major catechins, ranging in concentration from 21 to 107 pmol/gram tissue, could be detected in the prostate after the consumption of tea. By performing these types of studies, we can then begin to identify the pharmacological form(s) of key flavonoids and more accurately determine the amount that should be consumed in order to achieve the desired chemopreventive effect. Further, we may also better understand how inter-individual differences in the metabolic profiles of flavonoids contribute to their ultimate effect.

Chemical modifications of flavonoids have been shown to improve both their bioavailability and chemopreventive effects. In particular, substitution of hydroxyl moieties with methoxyl groups has been found to improve flavonoid bioavailability because the methoxyl groups are poor acceptors for the conjugating glucuronic or sulfate groups.[20] Thus, the methylated forms of apigenin and chrysin (5,7,4' trimethoxyflavone and 5,7' dimethoxyflavone, respectively) are more potent as compared to their hydroxylated forms, with respect to their ability to inhibit tumor cell growth in vitro. In addition, the methoxy-containing variants exhibit greater water solubility as compared to their hydroxyl-bearing counterparts. Similarly, *O*-methylation of chrysin has been shown to enhance its ability to inhibit the expression levels of pro-inflammatory cytokines in vitro.[21] However, at least with respect to anti-inflammatory activities, the site of methoxy substitution appears to impact relative potency, indicating a preference for an as yet, undefined structural conformation. Other chemical modifications that have been shown to enhance the bioavailability of flavonoids include the addition of sugar moieties (i.e., α-oligoglucosylation.).[22] The *in vivo* impact of these chemical modifications is yet to be determined.

Ongoing research efforts are also focused on using nanoparticles to improve the bioavailability of flavonoids.[23,24] These particles range in size from 1 to 100 nm. Nanoparticles typically consist of liposomes, emulsions, solid nanoparticles, poly(lactic-co-glycolic acid) nanoparticles, and micelles. They vary with respect to size, surface charge, size distribution, stability, and amount of material that is encapsulated. With respect to EGCG, nanoencapsulation has been found to significantly increase its stability and sustained release as compared to that of free EGCG. In rats or mice orally administered nanoencapsulated EGCG, a 1.5-fold higher plasma level of EGCG was observed. A greater than two-fold higher oral bioavailability resulted in an increased anti-tumor effect as compared to that of free EGCG. Similar studies have been performed using quercetin. Administration of nanoencapsulated quercetin to rats resulted in a five-fold increase in oral bioavailability, higher plasma levels of the parent compound, and a prolongation in its circulation as compared to that of free quercetin. In skin models of inflammation, nanoencapsulation of quercetin improved its absorption and ability to inhibit inflammation.

An additional approach that may be used is to maximize the amount of the most effective chemopreventive flavonoid that is produced within a given plant source or by a microorganism using bioengineering approaches. For example, a transgenic variety of flax with enhanced production of kaempferol, quercetin, and anthocyanins has been generated.[25] Alternatively, microorganisms such as *Escherichia coli* and *S. cerevisiae* may be used as the flavonoid source.[26] The primary advantage to using microorganisms versus plants to generate the desired flavonoid is their amenability to genetic manipulations.

Progress in understanding the relationship between flavonoids and the microbiome. Flavonoids and the microbiome participate in a mutually beneficial interaction, particularly within the gut.[27,28] The microbiome, via its ability to contribute toward flavonoid metabolism, can play an important role in the chemopreventive actions of flavonoids. For example, the extent to which flavonoids can be absorbed by traversing the intestinal epithelial barrier often depends on which metabolic products are formed and their ability to interact with transporters that guard the epithelial cell well. Metabolic conversion of flavonoids by the microbiome plays additional roles in determining the chemopreventive activity of flavonoids via their conversion to forms that are more potent with respect to their ability to impinge on receptors and signaling pathways. A significant amount of ingested flavonoids reaches the colon where the colonic microflora hydrolyzes and removes the conjugated moieties. The products of these reactions, the simple phenolic acids, can then be readily absorbed into the circulatory system. Use of an in vitro system of microbial degradation has revealed that amongst a variety of flavonoids, genistein, apigenin, kaempferol, and naringen, all of which are highly hydroxylated (5,7,4'-trihydroxyl flavonoids), were the most rapidly degraded. In human subjects, intestinal microflora has been shown to be a major contributor to the catabolism of tea catechins.[17] In addition, the intestinal flora metabolically generates kaempferol from isorhamnetin 3-O-glucoside.[29] Finally, the intestinal microflora exclusively converts the isoflavones, daidzein, and genistein into equol and 5-hydroxy-equol.[30] The microorganism involved here appears to be *S. isoflavoniconvertens*. This conversion has important biological implications as equol is a more potent activator of the estrogen receptors, as compared to that of the parent daidzein.[31] Interestingly, conversion of daidzein and genistein is affected by obesity which via its enhanced production of highly estrogenic agonists may account, in part, for the obesity-enhanced risk of breast cancer.[32]

Flavonoids in turn, impact the genetic population and function of the intestinal microbiome. For example, flavonoids have been shown to modulate the gut microbiome and in this manner suppress inflammation. Some flavonoids, in particular, the catechins found in tea, exert potent antimicrobial activities.[33] With respect to catechins, multiple mechanisms are involved. They can directly interact with the bacterial cell wall and damage the membrane. In addition, they can inhibit the function of essential bacterial enzymes such as those involved in fatty acid synthesis, tyrosine phosphorylation, DNA replication, and ATP synthesis. Catechin inhibition of viral function has also been reported. Flavonoid-induced changes in the bacterial population that colonizes the intestine have also been observed. Dietary administration of extracts prepared from apples engineered to express high levels of flavonoids (i.e., anthocyanins, epicatechin, procyanidin B2, and quercetin glycosides) resulted in mice with a higher bacterial load in the colon and higher numbers of *Bifidobacterium* spp. as compared to that of mice fed the control diets.[34] High numbers of *Bifidobacterium* spp. are typically associated with an anti-inflammatory effect. Similarly, dietary administration of extracts containing high levels of anthocyanins has also shown to increase the colon levels of *Bifidobacterium* spp. in both mice and in human subjects.[35] These results indicate that flavonoid-enriched food products may be useful as probiotics that would aid in restoring the gut microbiome of patients who suffer from inflammatory bowel disease (IBD) and other chronic disease states. They may also be useful in either preventing infections or ameliorating the oxidative stress and inflammatory conditions that accompany infections of *H. pylori*, hepatitis B, hepatitis C and other cancer-causing microorganisms. Thus, it is becoming increasingly clear that the microbiome plays an important role in the chemopreventive actions of flavonoids and may also be a significant contributor to the inter-individual response to their beneficial effects. Further research focused on understanding this microbiome-flavonoid interaction will aid in the development of approaches to be used for exploiting this relationship.

Lack of appropriate biomarkers to assess effective chemopreventive activities. While it is generally agreed that the patients most likely to benefit from chemopreventive approaches are those that are at high risk of developing cancer, the development and validation of biomarkers to be used for assessing their risk and establishing their diagnosis and prognosis are met with many challenges.[36] Further, the best marker to be used for determining whether or not flavonoid consumption or supplementation is effective in preventing the incidence or progression of cancer is unclear. Some studies indicate that the serum levels of pro-inflammatory cytokines may be appropriate.[14,15,37] Others argue that tumor expression of Ki-67, a marker of cell proliferation, is a good indicator, but that reproducibility between laboratories needs to be improved.[37] It would be more advantageous to use markers that are inexpensive to analyze and can be obtained from the patient using non-invasive techniques. Additional markers that are currently being explored include tumor-infiltrating natural killer cells, specific plasma microRNAs, intraepithelial neoplasia (identified using imaging) and PIK3CA (lung), PSA (prostate), or CA125/He4 (ovarian). It is highly likely that multiple markers will need to be identified, validated in pre-clinical models, and tested in large, randomized placebo controlled clinical trials.

Conclusions. The chemopreventive effects of flavonoids represent an as yet unrealized promise. However, significant strides have been made in overcoming the many hurdles that have been encountered and enhancing our understanding of how flavonoids exert their beneficial effects and how to best incorporate that knowledge toward improving human health. As we expand our basic understanding of cancer etiology, we will then be able to enhance our ability to accurately model these events in pre-clinical studies and perform rigorous and informative human population-based trials. Only then can this promise be realized.

References

1. Patterson SL, Colbert Maresso K, Hawk E. Cancer chemoprevention: Successes and failures. *Clin Chem.* Jan 2013;59(1):94–101.
2. Epelbaum R, Schaffer M, Vizel B, Badmaev V, Bar-Sela G. Curcumin and gemcitabine in patients with advanced pancreatic cancer. *Nutr Cancer.* 2010; 62(8):1137–1141.

3. Wheler J, Lee JJ, Kurzrock R. Unique molecular landscapes in cancer: Implications for individualized, curated drug combinations. *Cancer Res.* Dec 2014;74(24):7181–7184.

4. Barcellos-Hoff MH. Does microenvironment contribute to the etiology of estrogen receptor-negative breast cancer? *Clin Cancer Res.* Feb 2013;19(3): 541–548.

5. Nienhuis HH, Gaykema SB, Timmer-Bosscha H, *et al.* Targeting breast cancer through its microenvironment: Current status of preclinical and clinical research in finding relevant targets. *Pharmacol Ther.* Nov 2014.

6. Nichols HB, DeRoo LA, Scharf DR, Sandler DP. Risk-benefit profiles of women using tamoxifen for chemoprevention. *Journal of the National Cancer Institute.* Jan 2015;107(1):354.

7. Jordan VC. Avoiding the bad and enhancing the good of soy supplements in breast cancer. *Journal of the National Cancer Institute.* Sep 2014;106(9).

8. Chakrabarti A, Verbridge S, Stroock AD, Fischbach C, Varner JD. Multiscale models of breast cancer progression. *Ann Biomed Eng.* Nov 2012;40(11): 2488–2500.

9. Speroni L, Whitt GS, Xylas J, *et al.* Hormonal regulation of epithelial organization in a three-dimensional breast tissue culture model. *Tissue Engineering. Part C, Methods.* Jan 2014;20(1):42–51.

10. Yu M, Bardia A, Aceto N, *et al.* Cancer therapy: Ex vivo culture of circulating breast tumor cells for individualized testing of drug susceptibility. *Science.* Jul 2014;345(6193):216–220.

11. Ruggeri BA, Camp F, Miknyoczki S. Animal models of disease: Pre-clinical animal models of cancer and their applications and utility in drug discovery. *Biochem Pharmacol.* Jan 2014;87(1):150–161.

12. Harper CE, Patel BB, Wang J, Eltoum IA, Lamartiniere CA. Epigallocatechin-3-Gallate suppresses early stage, but not late stage prostate cancer in TRAMP mice: Mechanisms of action. *The Prostate.* Oct 2007;67(14):1576–1589.

13. El Touny LH, Banerjee PP. Identification of a biphasic role for genistein in the regulation of prostate cancer growth and metastasis. *Cancer Res.* Apr 2009;69(8):3695–3703.

14. Bobe G, Murphy G, Albert PS, *et al.* Serum cytokine concentrations, flavonol intake and colorectal adenoma recurrence in the Polyp Prevention Trial. *Br J Cancer.* Oct 2010;103(9):1453–1461.

15. Mentor-Marcel RA, Bobe G, Sardo C, *et al*. Plasma cytokines as potential response indicators to dietary freeze-dried black raspberries in colorectal cancer patients. *Nutr Cancer.* Aug 2012;64(6):820–825.
16. Terao J, Murota K, Kawai Y. Conjugated quercetin glucuronides as bioactive metabolites and precursors of aglycone *in vivo. Food and Function.* Jan 2011; 2(1):11–17.
17. Chow HH, Hakim IA. Pharmacokinetic and chemoprevention studies on tea in humans. *Pharmacological Research* (the official journal of the Italian Pharmacological Society). Aug 2011;64(2):105–112.
18. Lee J, Mitchell AE. Pharmacokinetics of quercetin absorption from apples and onions in healthy humans. *J Agric Food Chem.* Apr 2012;60(15): 3874–3881.
19. Lee MJ, Maliakal P, Chen L, *et al*. Pharmacokinetics of tea catechins after ingestion of green tea and (-)-epigallocatechin-3-gallate by humans: Formation of different metabolites and individual variability. *Cancer Epidemiol Biomarkers Prevent.* Oct 2002;11(10 Pt 1):1025–1032.
20. Walle T. Methylation of dietary flavones increases their metabolic stability and chemopreventive effects. *Int JMol Sci.* Nov 2009;10(11):5002–5019.
21. During A, Larondelle Y. The O-methylation of chrysin markedly improves its intestinal anti-inflammatory properties: Structure-activity relationships of flavones. *Biochem Pharmacol.* Dec 2013;86(12):1739–1746.
22. Murota K, Matsuda N, Kashino Y, *et al*. Alpha-Oligoglucosylation of a sugar moiety enhances the bioavailability of quercetin glucosides in humans. *Arch Biochem Biophys.* Sep 2010;501(1):91–97.
23. Khushnud T, Mousa SA. Potential role of naturally derived polyphenols and their nanotechnology delivery in cancer. *Molecular Biotechnology.* Sep 2013;55(1):78–86.
24. Wang S, Su R, Nie S, *et al*. Application of nanotechnology in improving bioavailability and bioactivity of diet-derived phytochemicals. *J Nutr Biochem.* Apr 2014;25(4):363–376.
25. Zuk M, Kulma A, Dyminska L, *et al*. Flavonoid engineering of flax potentiate its biotechnological application. *BMC Biotechnol.* 2011;11:10.
26. Trantas EA, Koffas MA, Xu P, Ververidis F. When plants produce not enough or at all: Metabolic engineering of flavonoids in microbial hosts. *Frontiers in Plant Science.* 2015;6:7.

27. Del Rio D, Rodriguez-Mateos A, Spencer JP, Tognolini M, Borges G, Crozier A. Dietary (poly)phenolics in human health: Structures, bioavailability, and evidence of protective effects against chronic diseases. *Antioxid Redox Signal*. May 2013;18(14):1818–1892.

28. Ahmed Nasef N, Mehta S, Ferguson LR. Dietary interactions with the bacterial sensing machinery in the intestine: The plant polyphenol case. *Frontiers in Genetics*. 2014;5:64.

29. Du LY, Zhao M, Xu J, *et al*. Analysis of the metabolites of isorhamnetin 3-O-glucoside produced by human intestinal flora in vitro by applying ultra-performance liquid chromatography/quadrupole time-of-flight mass spectrometry. *J Agric Food Chem*. Mar 2014;62(12):2489–2495.

30. Matthies A, Loh G, Blaut M, Braune A. Daidzein and genistein are converted to equol and 5-hydroxy-equol by human intestinal Slackia isoflavonicon-vertens in gnotobiotic rats. *J Nutr*. Jan 2012;142(1):40–46.

31. Muthyala RS, Ju YH, Sheng S, *et al*. Equol, a natural estrogenic metabolite from soy isoflavones: Convenient preparation and resolution of R- and S-equols and their differing binding and biological activity through estrogen receptors alpha and beta. *Bioorg Med Chem*. Mar 2004;12(6):1559–1567.

32. Frankenfeld CL, Atkinson C, Wahala K, Lampe JW. Obesity prevalence in relation to gut microbial environments capable of producing equol or O-desmethylangolensin from the isoflavone daidzein. *Eur J Clin Nutr*. Apr 2014;68(4):526–530.

33. Reygaert WC. The antimicrobial possibilities of green tea. *Frontiers in Microbiology*. 2014;5:434.

34. Espley RV, Butts CA, Laing WA, *et al*. Dietary flavonoids from modified apple reduce inflammation markers and modulate gut microbiota in mice. *J Nutr*. Feb 2014;144(2):146–154.

35. Faria A, Fernandes I, Norberto S, Mateus N, Calhau C. Interplay between anthocyanins and gut microbiota. *J Agric Food Chem*. Jul 2014;62(29): 6898–6902.

36. Cairns L. Eurocan Platform meeting: European recommendations for bio-marker-based chemoprevention trials. *Ecancermedicalscience*. 2014;8:488.

37. Peluso I, Miglio C, Morabito G, Ioannone F, Serafini M. Flavonoids and immune function in human: A systematic review. *Crit Rev Food Sci Nutr*. 2015;55(3):383–395.

Index

www.ingramcontent.com/pod-product-compliance
Lightning Source LLC
Chambersburg PA
CBHW050602190326

41458CB00007B/2148